NUMBER THEORY

LIBRARY OF MATHEMATICS

edited by

WALTER LEDERMANN

D.Sc., Ph.D., F.R.S.Ed., Professor of
Mathematics, University of Sussex

NUMBER THEORY

BY

T. H. JACKSON

Department of Mathematics
University of York

ROUTLEDGE & KEGAN PAUL

LONDON AND BOSTON

First published in 1975
by Routledge & Kegan Paul Ltd
Broadway House, 68–74 Carter Lane,
London EC4V 5EL and
9 Park Street,
Boston, Mass. 02108, USA
Set in Monophoto Times Roman
at the Universities Press, Belfast
and printed in Great Britain by
Unwin Brothers Ltd

ISBN 0 7100 7998 2

Cont

NUMBER THEORY

BY

T. H. JACKSON

Department of Mathematics
University of York

ROUTLEDGE & KEGAN PAUL

LONDON AND BOSTON

First published in 1975
by Routledge & Kegan Paul Ltd
Broadway House, 68–74 Carter Lane,
London EC4V 5EL and
9 Park Street,
Boston, Mass. 02108, USA
Set in Monophoto Times Roman
at the Universities Press, Belfast
and printed in Great Britain by
Unwin Brothers Ltd

ISBN 0 7100 7998 2

Contents

CONTENTS

Preface

The object of this book is to give an elementary introduction to the concepts and results of the theory of numbers. In a book of this size it is not possible to give a self-contained introduction to all the many branches of number theory, and so we concentrate on purely classical arguments. This has the advantage of requiring less mathematical background in the reader. At the same time elementary arguments bring out quite clearly the essential flavour of the subject, and make the book suitable for beginner students to read by themselves.

The first chapter deals with the fundamentals of divisibility, primality, and multiplicative functions. In Chapter 2 congruences are introduced, and the relationships between polynomial congruences and polynomial equations are emphasized. Chapter 3 discusses quadratic congruences in detail and the results obtained are used in Chapter 4 to study the representations of natural numbers by sums of squares.

I should like especially to thank the editor, Professor W. Ledermann, for his many constructive criticisms. I should also like to thank Dr A. M. Arthurs for his helpful advice; and Miss J. Houghton for her careful typing of my manuscript.

University of York   T. H. JACKSON

CHAPTER ONE

Basic Ideas

---

1. Introduction

Number theory shares with geometry the honour of being the oldest part of mathematics. It deals with whole or natural numbers, 1, 2, 3, 4, ..., which are used universally for counting and every other kind of calculation. Just as geometry has formed the basis of much modern mathematics, for example topology and analysis, so number theory has been the foundation of, and continues to stimulate, a great deal of modern algebra.

From earliest times until about three hundred years ago scholars attached mystical and magical properties to numbers, and numerology was a legitimate field of study. But nowadays we are interested in more basic properties arising from the ordinary concepts of addition and multiplication. In the work of Fermat (1601–65) the first great advances in number theory were made since the Greeks. Since his time other famous mathematicians, such as Euler (1707–83), Lagrange (1736–1813), Gauss (1777–1855), Hilbert (1862–1942) and Hardy (1878–1948), have introduced new and powerful methods, so that number theory has enjoyed a vigorous period of growth which continues today.

Number theory now encompasses several branches—classical, analytic, algebraic and the geometry of numbers.

1

This book is concerned mainly with classical results, and with modern methods for attacking classical problems. For example, the property that some numbers such as $5 = 1^2 + 2^2$ can be expressed as the sum of two square numbers, while others such as 7 cannot, will be examined. This is a problem first explored by Fermat. We shall use Gauss's theory of congruences to see why there are no natural numbers x and y such that $x^2 - 3y^2 = 2$. And we shall use Euclid's algorithm to show how to find natural numbers x and y which do satisfy the equation $11x - 13y = 1$. Indeed the whole problem of finding natural numbers which satisfy particular equations was one which greatly interested the Greeks.

A knowledge of set theory will not be required but it will be convenient to use some elementary set theory notation. We shall let $\mathbf{N}$ be the name of the collection or set of all natural numbers. For example:

$$a \in \mathbf{N},$$

which means that a is a member of $\mathbf{N}$. Sometimes it will be helpful to consider the set of all whole numbers $\ldots -3, -2, -1, 0, 1, 2, 3, 4, \ldots$, which are called integers. The set of all integers is denoted by $\mathbf{Z}$; thus if x is any integer:

$$x \in \mathbf{Z}.$$

A natural number therefore is simply a positive integer.

An important property of the integers is the **well-ordering principle** which states that any non-empty set of positive integers contains a smallest member. Indeed an equivalent form of this statement is that if each member of a given set of integers is greater than some number a, say, then the given set must contain a smallest member. For example, the set of even numbers, each of which is greater than -10, contains a least member, namely -8. We also have another basic property which again can be shown to be an alternative version of the same principle. That is, if each member of a given set of integers

2

is less than a fixed number b, say, then the given set of integers must contain a largest member. Thus the set of those integers each of which is less than 20, contains a largest member, namely 19. Although these are quite simple and evident properties of integers, they are not shared by every type of number system. Because, if we consider fractional numbers, the collection of all positive fractions does not contain a smallest positive fractional number. We shall see later that the well-ordering principle is often an essential aid in deducing further results about integers.

2. The elementary operations

If a and b are any two natural numbers (or integers), they have a sum, $a + b$, and a product, $a . b$ or ab, which are also natural numbers (integers). It will be assumed that the reader is familiar with the ordinary rules for calculating with addition and multiplication. Once we have addition and multiplication, it is easy to construct the inverse operations of subtraction and division.

Subtraction. If a and b are integers, their difference, $a - b$, is the integer d with the property that $a = b + d$.

The advantage of introducing the set **Z** of all integers is that any two integers a and b have a difference which is again an integer. It is not always possible to form the difference of two natural numbers with a result which is again a natural number. In practice, however, if a and b are given natural numbers—or indeed integers—it is usually easy to tell whether $a - b \in \mathbf{N}$; for this possibility just corresponds to the statement 'a is greater than b' (written $a > b$).

Division. If a and b are integers with $a \neq 0$ and $b = aq$ for some integer q, we say that a divides b, a is a divisor (or factor) of b, or b is a multiple of a. If a divides b we write $a \mid b$ and if a does not divide b we write $a \nmid b$.

If a and b are integers with $a \neq 0$, a may not be a divisor of b, and in contrast to subtraction, it may be quite difficult to decide whether a divides b. A large part of number theory is concerned with the study of divisibility (the concept of division) and the theories arising from it. All of the previously mentioned examples involve these theories for their solution, and this first chapter will be concerned entirely with the elementary properties of division. We shall begin our study of divisibility by using its definition to deduce two useful properties:

$$\text{if } a \mid b \text{ and } b \mid c \text{ then } a \mid c, \tag{1.1}$$

$$\text{if } a \mid b \text{ and } a \mid c \text{ then } a \mid (bx + cy) \text{ for any } x, y \in Z. \tag{1.2}$$

In order to see that (1.1) is true suppose that say $b = aq$ and $c = bm$ for some integers q and m. Then substituting for b we have $c = aqm$ so $a \mid c$. In (1.2) if $b = aq$ and $c = an$ then we have $bx + cy = a(qx + ny)$, so by definition $a \mid bx + cy$.

Note that from property (1.2), if a divides two numbers, it must also divide their sum and difference. For example, choosing $x = 1, y = -1$, (1.2) becomes

$$a \mid b \text{ and } a \mid c \text{ implies } a \mid b - c.$$

If we choose two natural numbers a and b it is possible that a will not divide b. However, we can still obtain useful information which relates a to b by applying the following process of 'division with remainder'.

Division with remainder. For each integer b and natural number a there are unique integers q and r such that

$$b = aq + r \text{ with } 0 \leqslant r < a. \tag{1.3}$$

First of all note that for a given a and b there are always some multiples of a which do not exceed b. Because, if b is positive or zero then $a \cdot 0 = 0 \leqslant b$, and if b is negative then $ab \leqslant b$.

So, using the well-ordering principle, we choose aq to be the largest multiple of a not exceeding b. That is,

$$aq \leqslant b \quad \text{and} \quad (q + 1)a > b, \qquad (1.4)$$

and we set $r = b - aq$. Then we have $b = aq + r$ as required and, from (1.4), $r = b - aq \geqslant 0$. Also from (1.4),

$$a - r = (q + 1)a - b > 0.$$

So we have established (1.3). In order to show that there is only one choice of q, r for given a, b, we suppose that we also have

$$b = aq_1 + r_1 \text{ with } 0 \leqslant r_1 < a \qquad (1.5)$$

and prove that $q_1 = q, r_1 = r$. We shall deduce that $q_1 = q$ by showing that either of the other two possibilities $q_1 < q$, $q_1 > q$ would contradict (1.3) and (1.5). So suppose if possible that $q_1 < q$. Then we must actually have $q_1 \leqslant q - 1$, since q_1 and q are both whole numbers. Using $q_1 \leqslant q - 1$ we see from (1.3) and (1.5) that

$$r_1 = b - aq_1 \geqslant b - a(q - 1) = b - aq + a = r + a \geqslant a.$$
$$(1.6)$$

But (1.5) says that $r_1 < a$, so (1.6), which says $r_1 \geqslant a$, is the desired contradiction. The possibility $q_1 > q$ similarly leads to the statement $r \geqslant a$ which contradicts (1.3). So in order to avoid a contradiction with (1.3) and (1.5) we must have $q_1 = q$. If we now replace q_1 by q in (1.5), and use (1.3), we see that

$$b = aq + r = aq + r_1$$

whence $r_1 = r$ as desired.

Equation (1.3) is just a formal statement of the familiar process of obtaining the remainder when one number is divided by another. For example, when $a = 5, b = 13$, (1.3) is true if we choose $q = 2$ and $r = 3$ because

$$13 = 5 \cdot 2 + 3.$$

Indeed, as far as 5 is concerned, there is a very close similarity between 3, 8, 13, 18, ... because they all leave the same remainder on division by 5. In Chapter 2 we shall see that this simple concept of leaving the same remainder can be systematically exploited to provide a very powerful tool in the theory of numbers. Note that statement (1.3) includes the possibility that $a \mid b$, for this is the case whenever $r = 0$.

In property (1.2) we encountered the idea of one number being a common divisor of two others. If a and b are two given numbers, an interesting problem is to investigate the sets of common multiples and common divisors of a and b.

3. Common multiples

If a and b are given natural numbers any number which is divisible by each of them is called a common multiple of a and b. There certainly are common multiples of any pair a, b, as the product ab is one. The smallest natural number which is a common multiple is called the **least common multiple** (l.c.m.) of a and b, sometimes written as $[a, b]$. It is singled out because every common multiple is divisible by the least one. In order to see this suppose that m is the l.c.m. of a and b and that M is any other common multiple of them. Then by the division with remainder property there are numbers q, r satisfying

$$M = qm + r \quad \text{and} \quad 0 \leqslant r < m. \qquad (1.7)$$

We shall show that $r = 0$. Now from (1.7) $r = M - qm$. Thus a must divide r, since it divides both m and M. Similarly b divides r, because it divides both m and M. So r is a common multiple of a and b, and from (1.7) $r < m$. Hence r cannot be positive as m is the least positive common multiple. Therefore $r = 0$ and $m \mid M$.

Example Take $a = 4$, $b = 6$. Their product is 24, which is a common multiple; but the least common multiple of a and b is 12. The set of all common multiples of 4 and 6 can therefore be characterized as the set of all multiples of 12.

4. Common divisors

If a and b are integers, not both zero, then any positive integer which divides both of them is called a common divisor of a and b. The largest such integer is called their greatest common divisor and is denoted by (a, b). There is always at least one common divisor of a and b, namely 1, and in some cases there may be only one. In any case each common divisor of a and b is a divisor of (a, b). This will appear as a consequence of the following theorem which is another application of division with remainder.

THEOREM   *Suppose a and b are non-zero integers and (a, b) is their greatest common divisor. Then there are integers x_0, y_0 such that*

$$(a, b) = ax_0 + by_0. \tag{1.8}$$

Remark   A multiple of a plus a multiple of b (such as in (1.8)) is said to be a **linear combination** of a and b.

Proof   Consider the set S of all those integers each of which can be expressed in the form $ax + by$, with integer coefficients x, y. Since $a = a \cdot 1 + b \cdot 0 \in S$, and $-a = a \cdot (-1) + b \cdot 0 \in S$, we see that S contains some positive members. Suppose the least positive integer in S is d; so that, in particular, $d = ax_0 + by_0$ for some integers x_0, y_0. We shall use division with remainder to show that d divides both a and b. Firstly, there are integers q, r such that

$$a = qd + r, 0 \leqslant r < d.$$

So $r = a - qd = a - q(ax_0 + by_0) = a(1 - qx_0) + b(-qy_0)$. Thus $r \in S$ and $r < d$. Since d is the least positive member of S, we must have $r = 0$ and $d \mid a$. Similarly $d \mid b$, so that d is a common divisor of a and b. Also, because $d = ax_0 + by_0$, it follows from the elementary properties of division that d must itself be divisible by every common divisor of a and b.

Therefore d is the greatest common divisor of a and b, so the theorem has been proved.

From the representation (1.8) it follows that (a, b) must be a multiple of every common divisor of a and b. The set of common divisors of a and b is therefore the set of all divisors of (a, b).

Example If $a = 24$, $b = 42$, then $(a, b) = 6$ and in accordance with (1.8) we have

$$6 = 24 \cdot 2 + 42 \cdot (-1).$$

Also we see that the common divisors of 24 and 42 are just the positive divisors of 6, namely 1, 2, 3, 6.

A particularly interesting and important situation arises when the only common divisor of a and b is 1 (which is then their greatest common divisor). **Whenever $(a, b) = 1$, a and b are said to be relatively prime.** If a and b are relatively prime we see from (1.8) that there are integers x_1, y_1 satisfying

$$ax_1 + by_1 = 1. \tag{1.9}$$

Example 7 and 12 are relatively prime and

$$12 \cdot 3 - 7 \cdot 5 = 1.$$

An important property of relatively prime numbers is that

if $(a, b) = 1$ and c is another number such that $a \mid bc$, then $a \mid c$.
$$\tag{1.10}$$

To see this, multiply both sides of (1.9) by c giving

$$acx_1 + bcy_1 = c.$$

Then a divides both ac and bc, so it must also divide c.

Example 1 If $(a, b) = 1$ and c is divisible by a and by b then it must be divisible by ab. This is because $a \mid c$ means $c = aq$ for some integer q. Then $b \mid c = aq$ and $(a, b) = 1$ imply $b \mid q$ so $q = bd$ for some integer d. Therefore $c = aq = abd$ or in other words $ab \mid c$. Indeed repeating this argument shows that if every two of the numbers $a_1, \ldots, a_n$ ($n \geqslant 2$), are relatively prime, and if c is divisible by each of the a's, then it must also be divisible by their product.

Example 2 If n is relatively prime to each of the numbers $a_1, \ldots, a_s$ then it must also be relatively prime to their product. That is,

$$(a_1, n) = (a_2, n) = \ldots = (a_s, n) = 1 \text{ implies } (a_1 a_2 \ldots a_s, n) = 1. \quad (1.11)$$

For we know that there must be pairs of integers $x_1, y_1, \ldots, x_s, y_s$ such that $a_1 x_1 + n y_1 = 1, \ldots, a_s x_s + n y_s = 1$. Multiplying together all these linear combinations* $a_i x_i + n y_i$ we obtain

$$(a_1 x_1 + n y_1)(a_2 x_2 + n y_2) \ldots (a_s x_s + n y_s) = a_1 \ldots a_s K + nL = 1,$$

where $K = x_1 x_2 \ldots x_s$ and $L = y_1 a_2 x_2 \ldots a_s x_s + \ldots + n^{s-1} y_1 \ldots y_s$ are integers. Since $(a_1 \ldots a_s, n)$ divides $a_1 \ldots a_s K + nL$, and so divides 1, we must have $(a_1 \ldots a_s, n) = 1$.

Suppose a and b are given natural numbers and we want to find the value of (a, b). We use our knowledge that (a, b) is the least positive integer in the set S of numbers of the form $ax + by$. We already know two members, a and b, of S and we use division with remainder repeatedly to find smaller numbers in S. The systematic procedure involved is known as **Euclid's Algorithm**. We may choose the names a, b so that $a \leqslant b$. Then if $a \mid b$ we have $a = (a, b)$. Otherwise, for some numbers q_1, r_1,

$$b = a q_1 + r_1 \quad \text{and} \quad 0 < r_1 < a. \quad (1.12)$$

We continue by repeatedly trying to divide the current divisor (in this case a) by the latest remainder (here r_1). We can do this as long as the latest remainder is not zero; so we obtain a sequence of equations as follows;

$$
\begin{aligned}
a &= r_1 q_2 + r_2 & 0 < r_2 < r_1 \\
r_1 &= r_2 q_3 + r_3 & 0 < r_3 < r_2 \\
&\cdots & \cdots \\
r_{n-2} &= r_{n-1} q_n + r_n & 0 < r_n < r_{n-1} \\
r_{n-1} &= r_n q_{n+1}.
\end{aligned}
\quad (1.13)
$$

Whatever the values of a and b the process must always

---

* As is usual the notation $a_i x_i + n y_i$ denotes the ith one of the expressions $a_1 x_1 + n y_1, \ldots, a_s x_s + n y_s$.

terminate as in (1.13). This is because the non-zero remainders are all different (each is less than the preceding one) and less than a, so there are at most $a - 1$ of them. We shall show that (a, b) is actually r_n, the last non-zero remainder obtained in the algorithm. Now the general equation occurring in (1.13) is

$$r_{i-1} = r_i q_{i+1} + r_{i+1}.$$

From this we deduce that (r_{i-1}, r_i) (which divides r_{i-1} and r_i) must divide r_{i+1}, so (r_{i-1}, r_i) divides (r_i, r_{i+1}). Similarly (r_i, r_{i+1}) divides r_{i-1} and so divides (r_{i-1}, r_i). Therefore $(r_{i-1}, r_i) = (r_i, r_{i+1})$; and since this reasoning holds for each of the equations in (1.12) and (1.13) we have

$$(a, b) = (a, r_1) = (r_1, r_2) = \ldots = (r_{n-2}, r_{n-1}) = (r_{n-1}, r_n) = r_n.$$

Example Use Euclid's algorithm to find the greatest common divisor of 130 and 741. Here the equations corresponding to (1.12) and (1.13) are

$$741 = 5 \cdot 130 + 91 \tag{1.14}$$
$$130 = 1 \cdot 91 + 39 \tag{1.15}$$
$$91 = 2 \cdot 39 + 13 \tag{1.16}$$
$$39 = 3 \cdot 13.$$

So 13 is the required greatest common divisor since it is the last non-zero remainder.

Euclid's algorithm can also be used if desired to find a representation of (a, b) in the form (1.8). In the above example (1.16) gives

$$13 = 91 - 2 \cdot 39$$
$$= 91 - 2(130 - 91) \text{ from (1.15)}$$
$$= 3 \cdot 91 - 2 \cdot 130 = 3(741 - 5 \cdot 130) - 2 \cdot 130 \text{ by (1.14)}$$
$$= 3 \cdot 741 - 17 \cdot 130.$$

This method is perfectly general, since the penultimate equation in (1.13) can always be written as

$$r_n = r_{n-2} - q_n r_{n-1}. \tag{1.17}$$

The earlier equations in (1.13) are then used successively to replace each remainder on the right hand side of (1.17) by a combination of two previous remainders. Eventually (1.12) is used to replace r_1 by $b - aq_1$ and $r_n = (a, b)$ is obtained as a linear combination of a and b.

Example Find integers x and y such that $11x - 13y = 1$. $(11, 13) = 1$ so from Euclid's algorithm we have

$$13 = 11 + 2$$
$$11 = 5 \cdot 2 + 1$$

so that

$$1 = 11 - 5 \cdot 2$$
$$= 11 - 5(13 - 11)$$
$$= 6 \cdot 11 - 5 \cdot 13.$$

Suitable values of x and y are therefore 6 and 5 respectively. Notice that these are not the only values of x and y which make $11x - 13y = 1$. We could, for example, take $x = 19$ and $y = 16$; or indeed $x = 6 + 13k$, $y = 5 + 11k$, where k is any integer, since

$$11(6 + 13k) - 13(5 + 11k) = 11 \cdot 6 - 13 \cdot 5 = 1.$$

5. Primes and prime factorization

Every natural number a, other than 1, has at least two positive divisors, namely 1 and a. Natural numbers which have exactly two positive divisors are called **prime numbers** or **primes**. For example 2, 3, 5, 7, and 11 are all primes. Natural numbers which are greater than 1 and which are not prime are said to be **composite** and by definition a composite number n has a factor a such that $1 < a < n$. For example $12 = 3 \cdot 4$, 15, 20 and 21 are composite. The number 1 is called a **unit** and is not classified as either prime or composite. The reason for the importance of prime numbers is given by the theorem below. It shows that primes are the building blocks from which all natural numbers are constructed by means of multiplication.

THEOREM *Every natural number other than 1 is either prime or is expressible as a product of primes.*

11

Proof We shall prove the theorem by induction. The result is true for 2, since 2 is prime. So it will be sufficient to suppose the theorem is true for all numbers less than n ($n > 2$) and show that it holds for n. If n is prime the statement is true. If n is composite then

$$n = a \cdot b \text{ where } 1 < a < n \text{ and } 1 < b < n.$$

By the inductive hypothesis a and b are each either prime or expressible as a product of primes. So suppose we have

$$a = a_1 \cdot a_2 \ldots a_k$$
$$b = b_1 \cdot \ldots b_l$$

where $a_1, \ldots, a_k, b_1, \ldots, b_l$ are primes and $k \geqslant 1, l \geqslant 1$. Then $n = a_1 \ldots a_k \cdot b_1 \ldots b_l$, and this is an expression for n as a product of primes.

A single prime number can be described as a product consisting of one term. A product having no terms (an empty product) is by convention given the value one. This convention, which is used throughout mathematics, is very helpful in simplifying enunciations of results. For instance, the statement of the theorem now becomes: **every natural number can be represented as a product of primes.** Indeed, as we see below, such a representation is unique apart from the order of the factors and the insertion of extra units into the product.

THE FUNDAMENTAL THEOREM OF ARITHMETIC *Every natural number can be represented in one and only one way as a product of primes (apart from the order of the factors and the number of units in the product).*

Proof The theorem is true for the number 2 so it will again be sufficient to assume that the result is true for all numbers less than n ($n > 2$) and show that it holds for n. We have already seen that n can be written as a product of primes, so suppose that this can be done in at least two ways:

$$n = p_1 \ldots p_r = q_1 \ldots q_s \qquad r \geqslant 1, s \geqslant 1, \qquad (1.18)$$

where the p's and q's denote prime numbers. If none of the q's were equal to p_1, then each q would be relatively prime* to p_1; so that, from (1.11), their product $q_1 q_2 \ldots q_s$ would also be relatively prime to p_1, contradicting (1.18). Therefore, by relabelling the q's if necessary, we can assume that p_1 is actually the first q, thus $p_1 = q_1$. Dividing throughout (1.18) by $p_1 = q_1$ we then obtain

$$\frac{n}{p_1} = p_2 \ldots p_r = q_2 \ldots q_s. \qquad (1.19)$$

Since $n/p_1 < n$, our assumption implies that $p_2 \ldots p_r$ is the unique representation of n/p_1 as a product of primes. Thus, from (1.19), $r = s$ and all the q's are p's (perhaps in some other order), so that n can be expressed as a product of primes in only one way apart from the order of the factors.

Examples When expressed as products of primes we have

    (i) $15 = 3 \cdot 5$ (ii) $100 = 2 \cdot 2 \cdot 5 \cdot 5$

   (iii) 521 is prime   (iv) $7007 = 7 \cdot 7 \cdot 11 \cdot 13$

If we compare the set of natural numbers to the set of all distinct chemical compounds, the prime numbers correspond to the atoms of individual elements. Just as we can form any molecule by combining appropriate elements, so we can form any number by combining appropriate prime numbers. However, this analogy is imperfect because there are examples of distinct molecules which are composed of the same atoms— the geometrical arrangement of the atoms in space being different. An equally significant difference though, is the fact, attributed to Euclid,† that **there are an infinite number of**

---

* The only divisors of p_1 are 1 and p_1 so we must have $(p_1, q) = 1$ or $(p_1, q) = p_1$. This last possibility is excluded if $p_1 \neq q$.

† (365–275 BC).

primes. This amounts to the statement that there is always a prime greater than any preassigned one. So suppose that P is a given prime number. We shall demonstrate the existence of a prime greater than P. Construct $N = 2 . 3 . 5 \ldots P + 1$ which is one more than the product of the prime numbers not exceeding P. None of the primes $2, \ldots, P$ divides N as they each leave remainder 1. On the other hand $N > 1$, so N is divisible by some prime, which is thus different from any of $2, \ldots, P$. This prime divisor of N is therefore the required prime number greater than P. Notice that in the above argument we could equally well have chosen N to be, say $(2 \ldots P) - 1$ or $(2 \ldots P)^2 + 1$. The only requirement is that N should be relatively prime to each of $2, \ldots, P$.

Now that we know there are an infinite number of primes, we are still left with the problem of deciding whether any given number is prime or not. There are factor tables* which give all the prime numbers less than 10,000,000 but for numbers above this limit the problem can be very difficult. We shall not discuss this question any further since the discovery of the greatest known prime numbers has required the use of large electronic computers (as well as powerful tests from the theory of numbers).

In the unique expression for a natural number n as a product of primes, several of the primes may be repeated (e.g. $12 = 2 . 2 . 3$). Suppose that $p_1, p_2, \ldots, p_r$ are the distinct primes occurring in the factorization of n, and that p_1 occurs a_1 times, p_2 occurs a_2 times, $\ldots$, and p_r occurs a_r times. Then we write

$$n = p_1^{a_1} \ldots p_r^{a_r}, \tag{1.20}$$

and the right hand side of (1.20) is called the canonical expression for n as a product of primes.

Example The canonical expressions of $12; 63; 360;$ are: $2^2 . 3; 3^2 . 7;$ and $2^3 . 3^2 . 5$ respectively.

* E.g. D. N. Lehmer, *Factor Table for the First Ten Millions*, Carnegie Institute publication no. 105.

When the canonical decomposition of a number into primes is known, its divisors can be easily classified. If $h \mid n$ and p is a prime occurring in the decomposition of h, then $p \mid n$ and from (1.20) we must have $p = p_i$ for some i with $1 \leqslant i \leqslant r$. Also from (1.20), p_i must not occur in h more than a_i times, so for h to be a divisor of n it is necessary and sufficient that

$$h = p_1^{\alpha_1} \ldots p_r^{\alpha_r}, \text{ where for each } i \, (1 \leqslant i \leqslant r) \, 0 \leqslant \alpha_i \leqslant a_i.$$

$$(1.21)$$

Example 1 Since $12 = 2^2 . 3$ the divisors of 12 are $2^0 3^0 = 1$, $2^1 3^0 = 2$, $2^2 3^0 = 4$, $2^0 3^1 = 3$, $2^1 3^1 = 6$, and $2^2 3^1 = 12$.

Example 2 Suppose m and n are any two natural numbers with, say, $n = p_1^{a_1} \ldots p_r^{a_r}$, $m = p_1^{b_1} \ldots p_r^{b_r}$. Then

$$(m, n) = p_1^{c_1} \ldots p_r^{c_r} \quad \text{and} \quad [m, n] = p_1^{d_1} \ldots p_r^{d_r},$$

where for each i $(1 \leqslant i \leqslant r)$, $c_i = \min (a_i, b_i)$ (the smaller of the two numbers a_i, b_i), and $d_i = \max (a_i, b_i)$. So for instance, if $n = 24 = 2^3 . 3^1 . 7^0$ and $m = 42 = 2^1 . 3^1 . 7^1$ then $(24, 42) = 2^1 . 3^1 . 7^0 = 6$ and $[24, 42] = 2^3 . 3^1 . 7^1 = 168$.

Example 3 Let n_1, n_2 be two relatively prime numbers. We can write them as $n_1 = p_1^{a_1} \ldots p_s^{a_s}$, $n_2 = p_{s+1}^{a_{s+1}} \ldots p_r^{a_r}$, for they cannot have a common prime factor since $(n_1, n_2) = 1$. Then $n_1 n_2 = p_1^{a_1} \ldots p_r^{a_r}$, so that each divisor h of $n_1 n_2$ can be written as $h = p_1^{\alpha_1} \ldots p_r^{\alpha_r} = h_1 h_2$ say, where $0 \leqslant \alpha_i \leqslant a_i$ for each i, and $h_1 = p_1^{\alpha_1} \ldots p_s^{\alpha_s}$, $h_2 = p_{s+1}^{\alpha_{s+1}} \ldots p_r^{\alpha_r}$. Thus each divisor h of $n_1 n_2$ can be expressed as the product of two relatively prime numbers h_1, h_2 which are divisors of n_1, n_2 respectively. Conversely if $h_1 \mid n_1$ and $h_2 \mid n_2$ then $h_1 h_2 \mid n_1 n_2$. Therefore as h_1 runs through all the divisors of n_1, and h_2 runs independently through all the divisors of n_2, the product $h_1 h_2$ goes without repetitions through all the divisors of $n_1 n_2$.

In (1.21) each different choice of exponents $\alpha_1, \ldots \alpha_r$ gives a different divisor of n (e.g. if each $\alpha_i = 0$ then $h = 1$; if each $\alpha_i = a_i$ then $h = n$). Since there are $(a_i + 1)$ choices for α_i, the total number of choices, (which is the number of distinct divisors of n) must be $(a_1 + 1)(a_2 + 1) \ldots (a_r + 1)$. We denote the number of divisors of n by $d(n)$ so we have:

$$d(n) = (a_1 + 1)(a_2 + 1) \ldots (a_r + 1). \qquad (1.22)$$

We can also use (1.21) to obtain the sum $\sigma(n)$ of the divisors of n. Indeed we have

$$\sigma(n) = (1 + p_1 + \ldots + p_1^{a_1})(1 + p_2 + \ldots + p_2^{a_2}) \ldots$$
$$\ldots (1 + p_r + \ldots + p_r^{a_r})$$
$$= \prod_{i=1}^{r} \frac{p_i^{a_i+1} - 1}{p_i - 1}, \qquad (1.23)$$

(where $\prod$ signifies a product). This is because multiplying out the expression $\prod_{i=1}^{r} (1 + p_i + p_i^2 + \ldots + p_i^{a_i})$ simply gives the sum of all possible products $p_1^{\alpha_1} \ldots p_r^{\alpha_r}$, where $0 \leqslant \alpha_i \leqslant a_i$ for each i. Now suppose that n_1 and n_2 are any two relatively prime numbers, with say

$$n_1 = p_1^{a_1} \ldots p_s^{a_s} \quad \text{and} \quad n_2 = p_{s+1}^{a_{s+1}} \ldots p_r^{a_r}.$$

Then

$$\sigma(n_1)\sigma(n_2) = \prod_{i=1}^{s} \frac{p_i^{a_i+1} - 1}{p_i - 1} \cdot \prod_{j=s+1}^{r} \frac{p_j^{a_j+1} - 1}{p_j - 1}$$
$$= \prod_{i=1}^{r} \frac{p_i^{a_i+1} - 1}{p_i - 1}$$
$$= \sigma(n_1 n_2).$$

So we have proved that the function σ has the following property,

$$(n_1, n_2) = 1 \text{ implies } \sigma(n_1 n_2) = \sigma(n_1)\sigma(n_2). \qquad (1.24)$$

We usually describe this property by saying that σ is a **multiplicative function.**

Examples   (i) $\sigma(p) = p + 1$ for a single prime p.
   (ii) $\sigma(2^k) = 2^{k+1} - 1$ for any natural number k.
   (iii) $\sigma(12) = \sigma(3)\sigma(2^2) = 4(2^3 - 1) = 4 \cdot 7 = 28$.

6. Multiplicative functions and the Möbius function

Any function $f(n)$, which is defined for each natural number n, is said to be multiplicative if

$$(n_1, n_2) = 1 \text{ implies } f(n_1 n_2) = f(n_1) f(n_2). \qquad (1.25)$$

Multiplicative functions occur frequently in number theory and have several interesting properties. For example, suppose that f is a multiplicative function and that the natural number n has the canonical decomposition $n = p_1^{a_1} \ldots p_r^{a_r}$. Then, since each prime power is relatively prime to all the others, we see, by applying (1.25) repeatedly, that

$$f(n) = f(p_1^{a_1}) \ldots f(p_r^{a_r}). \qquad (1.26)$$

So we need only evaluate f for prime power arguments in order to be able to determine its value for every natural number n.

We have already seen that the function $\sigma(n)$ is multiplicative. Another example is the function $d(n)$, since if $(n_1, n_2) = 1$ with, say, $n_1 = p_1^{\alpha_1} \ldots p_r^{\alpha_r}$, $n_2 = q_1^{\beta_1} \ldots q_s^{\beta_s}$, then

$$n_1 n_2 = p_1^{\alpha_1} \ldots p_r^{\alpha_r} \cdot q_1^{\beta_1} \ldots q_s^{\beta_s}$$

and

$$\begin{aligned} d(n_1 n_2) &= (\alpha_1 + 1) \ldots (\alpha_r + 1)(\beta_1 + 1) \ldots (\beta_s + 1) \\ &= d(n_1) \cdot d(n_2). \end{aligned}$$

Indeed, both $d(n)$ and $\sigma(n)$ have similar definitions as sums extended over the divisors of n, for we have

$$d(n) = \sum_{h \mid n} 1, \qquad \sigma(n) = \sum_{h \mid n} h, \qquad (1.27)$$

where for instance in the first sum above we include a 1 for each divisor of n, thus counting the total number of divisors. We can therefore simultaneously discuss both of these functions, and many others, by considering all functions $f(n)$

with the property that

$$f(n) = \sum_{h \mid n} g(h) \tag{1.28}$$

where $g(h)$ is some function of h alone. In the case of the two functions $f_1(n) = d(n)$ and $f_2(n) = \sigma(n)$, the corresponding functions $g_1(h)$ and $g_2(h)$ are given by $g_1(h) = 1$ for each h, and $g_2(h) = h$ for each h. It is easy to see that these functions g_1, g_2 are themselves multiplicative since if $(h, h') = 1$ (indeed if h, h' are any natural numbers) we have

$$g_1(h) \cdot g_1(h') = 1 \cdot 1 = 1 = g_1(hh')$$

and $$g_2(h) \cdot g_2(h') = hh' = g_2(hh').$$

The fact that the functions g_1, g_2 are multiplicative is actually sufficient to imply that the functions $d(n)$ and $\sigma(n)$ must be multiplicative, because we can prove in general that

if g is a multiplicative function, then the function f given by (1.28) is also multiplicative. (1.29)

This follows from the fact, which we already know, that if n_1 and n_2 are relatively prime the product $h_1 h_2$ ranges over all the divisors of $n_1 n_2$ if h_1 and h_2 range independently over the divisors of n_1 and n_2 respectively. Thus, if $(n_1, n_2) = 1$ and f and g are related by (1.28), we can write

$$f(n_1 n_2) = \sum_{h_1 \mid n_1, \, h_2 \mid n_2} g(h_1 h_2) = \sum_{h_1 \mid n_1, \, h_2 \mid n_2} g(h_1)g(h_2),$$

since g is multiplicative,

$$= \left(\sum_{h_1 \mid n_1} g(h_1) \right)\left(\sum_{h_2 \mid n_2} g(h_2) \right)$$

$$= f(n_1) \cdot f(n_2).$$

Example When $n_1 = 3$, $n_2 = 4$ we have $(n_1, n_2) = 1$, and the divisors of n_1 are 1, 3, while those of n_2 are 1, 2, 4. Therefore, in the particular case of the function σ,

$$\sigma(3 \cdot 4) = \sum_{h_1 \mid 3, \, h_2 \mid 4} h_1 h_2 = 1 \cdot 1 + 1 \cdot 2 + 1 \cdot 4 + 3 \cdot 1 + 3 \cdot 2 + 3 \cdot 4$$

$$= (1 + 3)(1 + 2 + 4)$$

$$= \left(\sum_{h_1 \mid 3} h_1 \right) \left(\sum_{h_2 \mid 4} h_2 \right)$$

$$= \sigma(3) \cdot \sigma(4).$$

Now suppose that $f(n)$ and $g(n)$ are two functions, defined for each natural number n, which are not necessarily multiplicative, but which satisfy the relation (1.28). Then, if we know the value of $g(h)$ for each natural number h, the given relation will determine the value of $f(n)$ for each natural number n. However, we can also use (1.28) to calculate each value of g in terms of those of f. This is because we can write (1.28) as

$$f(n) = g(n) + \sum_{h \mid n, \, h < n} g(h)$$

or

$$g(n) = f(n) - \sum_{h \mid n, \, h < n} g(h). \qquad (1.30)$$

So if for each $h < n$ we can write $g(h)$ in terms of values of f, then (1.30) enables us to do the same for $g(n)$. For example, from (1.28) we have

$f(1) = g(1),$ whence $g(1) = f(1),$

$f(2) = g(2) + g(1),$ $\qquad\qquad g(2) = f(2) - f(1),$

$f(3) = g(3) + g(1),$ $\qquad\qquad g(3) = f(3) - f(1),$

$f(4) = g(4) + g(2) + g(1),$ $\qquad g(4) = f(4) - f(2),$

$f(5) = g(5) + g(1),$ $\qquad\qquad g(5) = f(5) - f(1),$

$f(6) = g(6) + g(3) + g(2) + g(1),$ $\quad g(6) = f(6) - f(3) - f(2) + f(1).$

Actually, as the above values of $g(1)$ to $g(6)$ suggest, it is possible to give an explicit formula for $g(n)$ as a sum of the form

$$g(n) = \sum_{h \mid n} c_h f(h). \qquad (1.31)$$

19

In (1.31) each coefficient c_h may be either 0 or ± 1; and the value of every coefficient can be specified using Möbius's* function $\mu(a)$, which is defined for each natural number a by

$$\mu(a) = \begin{cases} 1 \text{ when } a = 1 \\ (-1)^r \text{ when } a \text{ is a product of } r \text{ distinct primes} \\ 0 \text{ when } a \text{ is divisible by the square of a prime.} \end{cases} \quad (1.32)$$

Examples   (i) $\mu(5) = -1$ (ii) $\mu(6) = (-1)^2 = +1$
                 (iii) $\mu(12) = 0$ (iv) $\mu(3 . 5 . 7) = (-1)^3 = -1$.

The function μ provides us with another example of a multiplicative function, because if a_1 and a_2 are relatively prime natural numbers we must have one of the following two situations.

(i) At least one of a_1, a_2, say a_1, is divisible by the square of a prime. Then $a_2 a_2$ is divisible by the square of the same prime so that $\mu(a_1 a_2) = 0 = 0 . \mu(a_2) = \mu(a_1) . \mu(a_2)$.

(ii) a_1 is a product of r distinct primes and a_2 is a product of s distinct primes, with $r \geqslant 0$ and $s \geqslant 0$ (where for instance $r = 0$ when $a_1 = 1$). Then $a_1 a_2$ is a product of $r + s$ distinct primes since a_1, a_2 cannot have a prime factor in common, and so $\mu(a_1 a_2) = (-1)^{r+s} = (-1)^r . (-1)^s = \mu(a_1) . \mu(a_2)$. Since μ is multiplicative we can deduce at once from (1.29) that the function $f_\mu(n) = \sum_{h \mid n} \mu(h)$ is also multiplicative. So in order to evaluate the sum $\sum_{h \mid n} \mu(h)$ for every natural number n, it suffices to determine its value when n is a prime power. We have:

$$f_\mu(1) = \sum_{h \mid 1} \mu(h) = \mu(1) = 1,$$

and when p is a prime and $\alpha \geqslant 1$,

$$f_\mu(p^\alpha) = \sum_{h \mid p^\alpha} \mu(h) = \mu(1) + \mu(p) + \ldots + \mu(p^\alpha)$$

$$= \mu(1) + \mu(p) = 0.$$

* 1790–1868.

Now each natural number $n > 1$ has a prime decomposition of the shape $p_1^{\alpha_1} \ldots p_r^{\alpha_r}$ where $r \geq 1$ and each exponent α_i is positive, whence $f_\mu(n) = f_\mu(p_1^{\alpha_1}) \ldots f_\mu(p_r^{\alpha_r}) = 0$. Therefore

$$\sum_{h \mid n} \mu(h) = \begin{cases} 1 \text{ when } n = 1 \\ 0 \text{ when } n > 1. \end{cases} \quad (1.33)$$

The importance of this result is the fact that if f and g are two functions which are related by (1.28), then (1.33) enables us to derive the inverse relationship

$$g(n) = \sum_{h \mid n} \mu\left(\frac{n}{h}\right) f(h) = \sum_{h' \mid n} \mu(h') f\left(\frac{n}{h'}\right). \quad (1.34)$$

The formulae in (1.34) are usually called the **Möbius inversion formulae.** Certainly the two displayed sums are the same, because as h runs through all the divisors of n, so also does $h' = \dfrac{n}{h}$. In order to show that both sums are equal to $g(n)$, we argue that from (1.28)

$$\begin{aligned}
\sum_{h \mid n} \mu(h) f\left(\frac{n}{h}\right) &= \sum_{h \mid n} \mu(h) \sum_{k \mid (n/h)} g(k) \\
&= \sum_{h \mid n} \sum_{k \mid (n/h)} \mu(h) g(k) \\
&= \sum_{hk \mid n} \mu(h) \cdot g(k) \\
&= \sum_{k \mid n} \sum_{h \mid (n/k)} g(k) \mu(h) \\
&= \sum_{k \mid n} g(k) \sum_{h \mid (n/k)} \mu(h).
\end{aligned}$$

From (1.33) we know that $\sum_{h \mid (n/k)} \mu(h)$ is 1 when $\dfrac{n}{k} = 1$ (i.e. when $k = n$), and is 0 otherwise. Thus the only non-zero term in the last double sum above is the term with $k = n$, which is $g(n) \cdot 1$. This completes the deduction of (1.34) from (1.28). It is also

21

possible to make the converse deduction of (1.28) from (1.34). That is, if f and g are two functions which satisfy the relation (1.34), then we can deduce from this that they must also satisfy the relation (1.28). This is because the sum $\sum_{h\mid n} g(h)$ on the right hand side of (1.28) can be expressed alternatively as $\sum_{h\mid n} g\left(\dfrac{n}{h}\right)$, and then, using (1.34), we have

$$\sum_{h\mid n} g\left(\frac{n}{h}\right) = \sum_{h\mid n} \sum_{k\mid (n/h)} \mu\left(\frac{n}{kh}\right) f(k)$$

$$= \sum_{hk\mid n} \mu\left(\frac{n}{kh}\right) f(k)$$

$$= \sum_{k\mid n} \sum_{h\mid (n/k)} f(k)\mu\left(\frac{n}{kh}\right)$$

$$= \sum_{k\mid n} f(k) \sum_{h\mid (n/k)} \mu\left(\frac{n}{kh}\right)$$

Again, from (1.33), the inner sum $\sum_{h\mid (n/k)} \mu\left(\dfrac{n}{kh}\right)$, which can also be written as $\sum_{h'\mid (n/k)} \mu(h')$, is 1 when $k = n$ and is 0 otherwise. Thus $\sum_{h\mid n} g\left(\dfrac{n}{h}\right) = f(n) \cdot 1$ as required.

Example Writing $d(n) = \sum_{h\mid n} g_1(h)$, the Möbius inversion formula gives $g_1(n) = \sum_{h\mid n} \mu(h)\, d\left(\dfrac{n}{h}\right)$. As every value of g_1 is 1, this last formula implies

$$1 = \sum_{h\mid n} \mu(h)\, d\left(\frac{n}{h}\right).$$

Similarly, from $\sigma(n) = \sum_{h\mid n} g_2(h)$, we have

$$g_2(n) = n = \sum_{h\mid n} \mu(h)\sigma\left(\frac{n}{h}\right).$$

We can also use the Möbius inversion formulae to deduce

22

other consequences whose statements do not involve the Möbius function. For example, we can now prove the following converse of property (1.29):

if the functions f and g are related by (1.28) and if f is multiplicative, then g must also be multiplicative. (1.35)

This is because (1.28) implies $g(n) = \sum_{h \mid n} \mu\left(\dfrac{n}{h}\right) f(h)$, whence, if $(n_1, n_2) = 1$,

$$g(n_1 n_2) = \sum_{h_1 \mid n_1,\, h_2 \mid n_2} \mu\left(\frac{n_1 n_2}{h_1 h_2}\right) f(h_1 h_2)$$

$$= \sum_{h_1 \mid n_1,\, h_2 \mid n_2} \mu\left(\frac{n_1}{h_1}\right) \mu\left(\frac{n_2}{h_2}\right) f(h_1) f(h_2)$$

$$= \left(\sum_{h_1 \mid n_1} \mu\left(\frac{n_1}{h_1}\right) f(h_1) \right) \left(\sum_{h_2 \mid n_2} \mu\left(\frac{n_2}{h_2}\right) f(h_2) \right)$$

$$= g(n_1) \cdot g(n_2).$$

7. Perfect and amicable numbers

In this section we shall discuss two classical problems concerned with $\sigma(n)$ which have still not been completely settled. They both arise from the great importance which medieval and earlier mathematicians attached to the divisors of a number n, and in particular to the **proper divisors** of n. Proper divisors of n are simply those divisors which are less than n itself, so that if $\sigma_0(n)$ is the sum of the proper divisors of n we have

$$\sigma(n) = \sigma_0(n) + n.$$

Each natural number n can then be classified as either deficient, perfect or abundant according as $\sigma_0(n)$ is respectively less than, equal to, or greater than n.

Examples   (i) 8 is deficient since $\sigma_0(8) = 1 + 2 + 4 = 7 < 8$
              (ii) 6 is perfect since $\sigma_0(6) = 6$
              (iii) 12 is abundant since $\sigma_0(12) = 16 > 12$.

Most numbers are either deficient or abundant. Perfect numbers are quite rare and people of every century have spent a great deal of effort trying to find them. The first real progress in this search was made when Euclid found a formula which produces even perfect numbers:

if $2^p - 1$ is prime for some p, then the number n given by

$$n = 2^{p-1} (2^p - 1), \qquad (1.36)$$

is perfect.

It is easy to verify this statement using our knowledge about $\sigma(n)$ and the fact that n is perfect means $\sigma(n) = 2n$ (since $\sigma(n) = \sigma_0(n) + n$). If n is given by (1.36) we have

$$\sigma(n) = \sigma(2^{p-1}) \cdot \sigma(2^p - 1) \text{ using (1.24)}$$
$$= (2^p - 1) \cdot 2^p \text{ (since } 2^p - 1 \text{ is prime)}$$
$$= 2n$$

so n is perfect. The fact that all even perfect numbers must be given by (1.36) was established much later by Euler with the following argument. Suppose we are given an even perfect number n and suppose, as we may, that $n = 2^{l-1} \cdot k$ where k is odd and $l > 1$. We shall show that k is prime and that $k = 2^l - 1$. We have,

$$\sigma(n) = \sigma(2^{l-1}) \cdot \sigma(k)$$
$$= (2^l - 1)\sigma(k)$$
$$= 2^l \cdot k \text{ since } n \text{ is perfect.}$$

That is

$$(2^l - 1)(\sigma_0(k) + k) = 2^l \cdot k,$$

or

$$(2^l - 1)\sigma_0(k) = k.$$

This last equation shows that $\sigma_0(k)$, the sum of the proper divisors of k, is itself a proper divisor of k. Thus $\sigma_0(k)$ must be the only proper divisor and so must be 1. Putting $\sigma_0(k) = 1$ we have $k = 2^l - 1$, and k must also be prime since 1 is its only proper divisor.

The above results of Euclid and Euler leave two major questions which modern mathematicians have been unable to answer:

(I) Are there any odd perfect numbers?

(II) Are there an infinite number of even perfect numbers?

The most that is known about (I) is that any odd perfect number would have to satisfy several stringent conditions and would have to be extremely large, so it is quite likely that there are no such numbers. The difficulty of question (II) lies in finding numbers p such that $2^p - 1$ is prime. There are at present 24 known suitable values of p ranging from $p = 2$ to $p = 19,937$. Apart from the fact that such numbers p have themselves to be primes the known examples show no discernible pattern, so the solution of question (II) still seems very distant.

Allied to the concept of a perfect number is that of a pair of **amicable numbers,** say m and n, which have the property that

$$\sigma_0(m) = n \quad \text{and} \quad \sigma_0(n) = m. \qquad (1.37)$$

Example   220 and 284 are amicable numbers since

$$\sigma_0(220) = 284 \quad \text{and} \quad \sigma_0(284) = 220.$$

Amicable numbers are more difficult to classify than perfect numbers, since there are several different formulae which can be used to produce amicable pairs. One such rule, due to Thabit ben Korrah,* is: if $p = 3 . 2^n - 1$, $q = 3 . 2^{n-1} - 1$ and $r = 9 . 2^{2n-1} - 1$ are all prime for some value of n greater than 1, then $2^n pq$ and $2^n . r$ are amicable numbers. For instance, when $n = 2$; $p = 11$, $q = 5$ and $r = 71$ are all primes so Thabit's rule gives the classical pair $2^2 . 5 . 11 = 220$ and $2^2 . 71 = 284$. Unfortunately there are only three known values of n (namely $n = 2, 4, 7$) for which Thabit's formula

---

* Ninth century.

gives amicable pairs. Indeed, although many amicable numbers have been found, it is still not known whether or not there are an infinite number of them.

EXERCISES

1. If m and n are any two natural numbers, prove that $(m, n) \cdot [m, n] = mn$.
2. Find the greatest common divisors of the following pairs of numbers a and b:

   (i) 67, 24;   (ii) 15, 72;   (iii) 36, 55;   (iv) 35, 126.

3. In each of the cases of question 2, express (a, b) as a linear combination of a and b.
4. Suppose that $a_1, \ldots, a_n$ are n natural numbers, and that their greatest common divisor is denoted by $(a_1, \ldots, a_n)$. Show that $(a_1, \ldots, a_n) \geqslant (a_1, (a_2, \ldots, a_n))$. Use this result to deduce by induction that $(a_1, \ldots, a_n)$ can be expressed as $a_1 x_1 + \ldots + a_n x_n$ for some integers $x_1, \ldots, x_n$, and that actually $(a_1, \ldots, a_n) = (a_1, (a_2, \ldots, a_n))$.
5. Find the greatest common divisors of the following sets of numbers:

   (i) 6, 15, 80;   (ii) 63, 105, 189, 287;   (iii) 45, 36, 60;   (iv) 19, 42, 51.

6. Show that if a number n is composite, it must have a prime divisor less than or equal to $\sqrt{n}$. [This means that in order to test whether a given number n is prime or composite, it is sufficient to test n for divisibility by each prime less than or equal to $\sqrt{n}$].
7. Use the result of question 6 to determine all the prime numbers less than 100.
8. Determine the canonical expressions, as products of primes, of the following numbers:

   (i) 130   (ii) 1386   (iii) 247   (iv) 307   (v) 1001.

9. Show that each odd prime is either one more than a multiple of 4, or one less than a multiple of 4.
10. Show that a product of numbers, each of which is one more than a multiple of 4, must also be one more than a multiple of 4. Deduce that a natural number which is one less than a multiple of 4 must have at least one prime divisor of the same form. Hence prove that there are infinitely many primes of the form $4k - 1$ by considering the divisors of $N = 4P_1 \ldots P_r - 1$, where each P_i is a prime of the shape $4k_i - 1$.
11. Prove, by a similar argument to that used in question 10, that there are infinitely many primes of the shape $6k - 1$.

12. Let $\sigma^k(n)$ denote the sum of the k-th powers of the divisors of $n = p_1^{\alpha_1} \ldots p_r^{\alpha_r}$; that is, $\sigma^k(n) = \sum_{h|n} h^k$. Prove that $\sigma^k(n) = \prod_{i=1}^{r} \left(\frac{p_i^{(\alpha_i + 1)k} - 1}{p_i^k - 1} \right)$, and use this formula to show that σ^k is multiplicative.

13. Prove that if $n = p_1^{\alpha_1} \ldots p_r^{\alpha_r}$ then $\sum_{h|n} |\mu(h)| = 2^r$.

CHAPTER TWO

Congruences and Equations

---

1. Congruences and complete sets of residues

If a, b are integers and m is a natural number, we shall say that **a and b are congruent with respect to the modulus m** if they both leave the same remainder when divided by m. In other words a and b are congruent with respect to m (or 'congruent modulo m') whenever

$$m \mid a - b, \tag{2.1}$$

and we describe this by writing

$$a \equiv b \pmod{m}. \tag{2.2}$$

If a is not congruent to b we write $a \not\equiv b \pmod{m}$.

Examples (i) $3 \equiv 8 \pmod{5}$ (ii) $-1 \equiv 13 \pmod{7}$
(iii) $2 \equiv 42 \pmod{8}$ (iv) $a \equiv a \pmod{m}$ and $m \equiv 0$
(v) $a \equiv b \pmod{1}$ for $\pmod{m}$ for any m
any integers a, b.

If we are given several congruences with respect to the same modulus, they can be combined together to produce new congruences in a fairly simple way. For example if $a_1 \equiv b_1 \pmod{m}$ and $a_2 \equiv b_2 \pmod{m}$, then

$$a_1 + a_2 \equiv b_1 + b_2 \pmod{m} \quad \text{and} \quad a_1 a_2 \equiv b_1 b_2 \pmod{m}. \tag{2.3}$$

This is because if m divides $a_1 - b_1$ and $a_2 - b_2$ it must also divide

$$(a_1 - b_1) + (a_2 - b_2) = (a_1 + a_2) - (b_1 + b_2)$$

and $$(a_1 - b_1)a_2 + b_1(a_2 - b_2) = a_1a_2 - b_1b_2.$$

We can now similarly combine a third congruence, say $a_3 \equiv b_3 \pmod{m}$, with those given by (2.3). Repeating the process shows that from the n congruences $a_1 \equiv b_1 \pmod{m}, \ldots, a_n \equiv b_n \pmod{m}$ we can conclude

$$a_1 + a_2 + \ldots + a_n \equiv b_1 + \ldots + b_n \pmod{m},$$

and $$a_1a_2 \ldots a_n \equiv b_1b_2 \ldots b_n \pmod{m}. \tag{2.4}$$

A useful deduction from (2.3) is that if we combine the congruences $a \equiv b \pmod{m}$ and $k \equiv k \pmod{m}$, where k is any integer, we obtain

$$ka \equiv kb \pmod{m}. \tag{2.5}$$

Also, if we combine the congruence $a \equiv b \pmod{m}$ with itself n times, as in (2.4), we have

$$a^n \equiv b^n \pmod{m}. \tag{2.6}$$

Although it is always valid to multiply both sides of a congruence by the same integer, or raise both sides to the same power, it is not always correct to divide both sides of a congruence by a common factor. For instance, in the congruence $6 \equiv 16 \pmod{10}$ it is not correct to divide both sides by 2 since $3 \not\equiv 8 \pmod{10}$. However, if in (2.2) d divides a, b and m, then

$$d \cdot \frac{m}{d} \,\bigg|\, d\left(\frac{a}{d} - \frac{b}{d}\right)$$

so $$\frac{m}{d} \,\bigg|\, \left(\frac{a}{d} - \frac{b}{d}\right).$$

That is,

$$\frac{a}{d} \equiv \frac{b}{d} \left(\bmod \frac{m}{d}\right) \text{ if } d \text{ is a common divisor of } a, b \text{ and } m. \quad (2.7)$$

Also, if d divides both a and b and is relatively prime to m then

$$m \mid d\left(\frac{a}{d} - \frac{b}{d}\right)$$

so

$$m \mid \left(\frac{a}{d} - \frac{b}{d}\right).$$

That is, $\qquad \dfrac{a}{d} \equiv \dfrac{b}{d} \,(\bmod\, m) \quad \text{if} \quad (d, m) = 1. \qquad (2.8)$

Now suppose that $c_n x^n + c_{n-1} x^{n-1} + \ldots + c_1 x + c_0 = f(x)$, say, is any polynomial with integer coefficients. If $a \equiv b \,(\bmod\, m)$, the addition and multiplication properties (2.4), (2.5), (2.6) give

$$c_n a^n \equiv c_n b^n \,(\bmod\, m)$$
$$\ldots$$
$$c_1 a \equiv c_1 b \,(\bmod\, m)$$
$$c_0 \equiv c_0 \,(\bmod\, m),$$

whence

$$c_n a^n + c_{n-1} a^{n-1} + \ldots + c_0$$
$$\equiv c_n b^n + c_{n-1} b^{n-1} + \ldots + c_0 \,(\bmod\, m).$$

In other words, **$a \equiv b \pmod{m}$ implies $f(a) \equiv f(b) \pmod{m}$.**

Example (Rule for divisibility by 9)
Suppose N is a natural number which can be written in ordinary decimal notation as $c_n c_{n-1} \ldots c_0$. That is,

$$N = c_n 10^n + c_{n-1} 10^{n-1} + \ldots + c_0.$$

Since $10 \equiv 1 \,(\bmod\, 9)$ we have

$$N \equiv c_n + c_{n-1} + \ldots + c_0 \,(\bmod\, 9),$$

so N is divisible by 9 (i.e. $N \equiv 0 \,(\bmod\, 9)$ if and only if the sum of its digits is divisible by 9.

If m is a given modulus and a is any integer, there will be integers q and r so that

$$a = qm + r \quad \text{and} \quad 0 \leqslant r < m.$$

Thus $m \mid a - r$; that is $a \equiv r \pmod{m}$. For a particular a there will only be one suitable r with $0 \leqslant r \leqslant m - 1$, so the congruence $a \equiv r \pmod{m}$ says that each integer is congruent $\pmod{m}$ to one and only one of the numbers $0, 1, \ldots, m - 1$. Any set of m integers which, on division by m, leaves all the remainders $0, \ldots, m - 1$, is called a **complete set of residues modulo m**. Since the numbers in such a set include all possible remainders, they have the property that each integer is congruent to one of them.

Examples The following sets are all complete sets of residues modulo 5:
   (i) $0, 1, 2, 3, 4$;   (ii) $-2, -1, 0, 1, 2$;   (iii) $-7, 1, 5, 9, 17$.

2. Congruences and equations

If, as before, $f(x)$ is a polynomial in x with integer coefficients, a common problem is to find the values of x which satisfy the equation

$$f(x) = 0. \tag{2.9}$$

Indeed, a number-theorist would ask: 'Are there any *integers x* which satisfy equation (2.9)?' When we are studying congruences modulo m, we can (by analogy with (2.9)) seek integer values of x which satisfy the **polynomial congruence**

$$f(x) \equiv 0 \pmod{m}. \tag{2.10}$$

Suppose that a particular integer a is a solution of (2.10)—that is, $f(a) \equiv 0 \pmod{m}$. Then we know that any integer b, congruent to a modulo m, will also be a solution of (2.10) (since $a \equiv b \pmod{m}$ implies $f(a) \equiv f(b) \pmod{m}$). So in order to see if there are any integers which satisfy (2.10) it is sufficient to test each of the integers in any complete set of residues modulo m. This will involve m trials, so the work can become quite tedious if either the degree of the polynomial or the modulus m is large. We shall see later that there are powerful theorems which can often be used to simplify the search for solutions of a polynomial congruence.

One straightforward simplification is to try to replace $f(x)$ in (2.10) by a simpler polynomial $g(x)$ and then solve the congruence $g(x) \equiv 0 \pmod{m}$. We can certainly replace $f(x)$ by $g(x)$ in (2.10) if $f(a) \equiv g(a) \pmod{m}$ for every integer a; for then $g(x) \equiv 0 \pmod{m}$ will have exactly the same solutions as the original congruence $f(x) \equiv 0 \pmod{m}$. If $f(x)$ and $g(x)$ satisfy this condition $f(a) \equiv g(a) \pmod{m}$ for every a, we say that $f(x)$ and $g(x)$ are identically congruent modulo m and write $f(x) \equiv g(x) \pmod{m}$. Notice that we can be sure of having $f(x) \equiv g(x) \pmod{m}$ if their corresponding coefficients are congruent modulo m; e.g. $x^3 - 12x^2 - 5x + 1 \equiv x^3 + x + 1 \pmod{6}$. This last observation is frequently very helpful.

The connection between equation (2.9) and the corresponding congruence (2.10) is really very close, since knowing the solutions of either one of them gives information about the solutions of the other. This is because if x_0 is an integer which is a solution of (2.9) (so $f(x_0) = 0$), then we certainly have $f(x_0) \equiv 0 \pmod{m}$ for any m, and thus x_0 is also a solution of (2.10). In other words **the solutions of (2.9) are all to be found among the solutions of (2.10).** For example, if we know that $x = a$ is the only solution of (2.10) in a particular complete set of residues, then in seeking solutions of equation (2.9) we may restrict our attention to integers which are congruent to a modulo m (since these are the only integers which satisfy congruence (2.10)). We can also see that **when there are no solutions to the congruence (2.10) there can be no solutions to the corresponding equation (2.9).**

Example 1　The equation $x^2 = 1$ has the two solutions $x = \pm 1$. These are therefore solutions of the congruence $x^2 \equiv 1 \pmod{8}$, as are all the integers congruent to $\pm 1 \pmod{8}$. The congruence, however, has the further solutions $x \equiv \pm 3 \pmod{8}$.

Example 2 Consider $f(x) \equiv 0 \pmod{11}$ where $f(x) = x^2 - 24x + 143$. We have $f(x) \equiv x^2 - 2x = x(x - 2) \pmod{11}$, so $x = 0$ and $x = 2$ are solutions of the congruence. If there are any solutions of the equation $f(x) = 0$, they must be found among the integers which are congruent to 0 or 2 modulo 11. In this case $x = 11$ and $x = 13$ are both solutions of $f(x) = 0$.

Example 3 The congruence $f(x) = x^2 - 6x + 11 \equiv 0 \pmod 6$ has the two solutions $x \equiv \pm 1 \pmod 6$. The integers congruent to ± 1 modulo 6 are therefore the only numbers which could possibly satisfy the equation $f(x) = 0$. In this case the equation actually has no solutions since $f(x) = (x - 3)^2 + 2$, which is always at least 2.

Example 4 The congruence $f(x) = 7x^4 - 35x^3 + x^2 - 3 \equiv 0 \pmod 7$ has no solutions. This is because $f(x) \equiv x^2 - 3 \pmod 7$ and $x^2 - 3 \equiv 0 \pmod 7$ has no solutions in the complete set of residues specified by $-3 \leqslant x \leqslant 3$. There are thus no integer values of x which satisfy the corresponding equation $f(x) = 0$.

The above relationships between a congruence and the corresponding equation still hold true if, in place of the polynomial $f(x)$, we have any polynomial (with integer coefficients) involving several variables, say $x_1, x_2, \ldots, x_n$. We denote a polynomial involving all these variables by $f(x_1, \ldots, x_n)$; for example, $f(x_1, x_2) = x_1^2 - 3x_1x_2 + x_2^2$ or $f(x_1, x_2, x_3) = x_1 + 5x_2^2x_3 - 6$. If $x_1 = a$, $x_2 = b$, $\ldots$ is a particular solution of the equation $f(x_1, \ldots, x_n) = 0$, that is, $f(a, b, \ldots) = 0$, then we shall certainly have $f(a, b, \ldots) \equiv 0 \pmod m$, so as before **the solutions of the equation will all be found among the solutions of the corresponding congruence.** Again, as a particular case, we see that **if the congruence $f(x_1, \ldots, x_n) \equiv 0 \pmod m$ has no solutions, then the equation $f(x_1, \ldots, x_n) = 0$ has no solutions.**

Example One of the equations mentioned at the beginning of the book was $x^2 - 3y^2 = 2$. Consider the corresponding congruence

$$x^2 - 3y^2 \equiv 2 \pmod 3.$$

Since $x^2 - 3y^2 \equiv x^2 \pmod 3$, we must look for solutions of the congruence $x^2 \equiv 2 \pmod 3$. Now it is easy to see that the squares of the numbers in any complete set of residues modulo 3 are each congruent to either 0 or 1. So in this case the congruence has no solutions. The original equation $x^2 - 3y^2 = 2$ has therefore no solutions in integers x, y.

In the rest of this chapter, we shall discuss polynomial congruences in one variable, and in the course of our investigations we shall come across some new ideas which are interesting and important in their own right.

3. Linear congruences

We shall first consider **linear congruences,** which are those where the polynomial $f(x)$ is of degree one. If $f(x) = ax - b$ where a ($\neq 0$) and b are integers, then the congruence $f(x) \equiv 0$ (mod m) is

$$ax \equiv b \,(\text{mod } m). \tag{2.11}$$

Solving the congruence (2.11) is the same as finding integers x and y which satisfy the equation $ax = b + my$, or

$$ax - my = b. \tag{2.12}$$

Each common divisor of a and m divides the left hand side of (2.12), and so must divide b. Thus for (2.12) to be soluble it is necessary that $(a, m) \mid b$. We shall show that this condition is also sufficient to ensure that (2.12) (and so (2.11)) has a solution. Suppose, then, that $(a, m) = d$ and $b = b'd$ where b' is an integer. From our result in equation (1.8) about greatest common divisors, we know that there are integers x', y' such that

$$ax' - my' = d.$$

Multiplying throughout by b', we see that the integers $x = b'x'$ and $y = b'y'$ satisfy

$$ax - my = b'd = b.$$

So **the necessary and sufficient condition for the congruence (2.11) to have a solution is that (a, m) $\mid$ b.** In particular if a and m are relatively prime ($(a, m) = 1$), then for any integer b the congruence $ax \equiv b$ (mod m) is soluble. Note that when a and m are relatively prime the congruence (2.11) has exactly one solution in any complete set of residues. For if x_1 and x_2 are both solutions of (2.11) then $ax_1 \equiv ax_2 \pmod{m}$. Thus $m \mid a(x_1 - x_2)$, whence $m \mid x_1 - x_2$ since $(a, m) = 1$, or in other words $x_1 \equiv x_2 \pmod{m}$.

Given any particular linear congruence it is now an easy matter to decide whether or not the congruence has a solution, but to actually find the solution (or solutions), it will usually still be necessary to test all the numbers in a complete set of residues.

Examples
(i) $4x \equiv 3 \pmod{6}$ is not soluble since $(4, 6) = 2$ and $2 \nmid 3$.
(ii) $2x \equiv 10 \pmod{8}$ is soluble since $(2, 8) = 2$ and $2 \mid 10$. By testing all the numbers in a complete set of residues we find the two solutions $x \equiv 1 \pmod{8}$ and $x \equiv 5 \pmod{8}$.
(iii) $3x \equiv 5 \pmod{11}$ is soluble since $(3, 11) = 1$. Because $5 \equiv 27 \pmod{11}$ we can write the congruence as $3x \equiv 27 \pmod{11}$, which on division by 3 gives the solution $x \equiv 9 \pmod{11}$.

If the polynomial $f(x)$ is of degree two or more, it becomes difficult to obtain simple conditions for the solubility of $f(x) \equiv 0 \pmod{m}$. We shall discuss these cases in Chapter 3, but first we shall show that for a prime modulus m we need only consider polynomial congruences where the polynomial in question is of degree less than m.

4. Reduced sets of residues and Euler's φ-function

Suppose now that a is any integer relatively prime to m. As usual we can find integers q and r so that

$$a = qm + r \quad \text{with} \quad 0 \leqslant r \leqslant m - 1;$$

and here $a \equiv r \pmod{m}$. Also, since $(a, m) = 1$, we have $(r, m) = 1$, because any common divisor of r and m would have to divide a and so would be a common divisor of a and m. Every integer relatively prime to m is therefore congruent to exactly one of the numbers in the set R of integers relatively prime to m and lying between 0 and $m - 1$ inclusive. The

CONGRUENCES AND EQUATIONS

number of members of R depends only on m and is usually denoted by $\phi(m)$. We shall call the numbers in R, $r_1, r_2, \ldots, r_{\phi(m)}$. Any set of $\phi(m)$ integers which, on division by m, leaves all the relatively prime remainders $r_1, \ldots, r_{\phi(m)}$, is called a **reduced set of residues modulo m.** Since the numbers in such a set include all the possible relatively prime remainders, they have the property that each integer, relatively prime to m, is congruent to one of them. Note that when $m > 1$, zero can never be a relatively prime remainder, as $(0, m) = m > 1$.

Examples
(i) $\phi(1) = 1$ since 0 is the only relatively prime remainder in this case.
(ii) $\phi(12) = 4$ because the following sets of four integers are all reduced sets of residues modulo 12:

$$\{1, 5, 7, 11\}, \qquad \{-13, 1, 17, 31\}, \qquad \{-31, -17, -1, 13\}.$$

(iii) $\phi(7) = 6$ since $\{1, 2, 3, 4, 5, 6\}$ is a reduced set of residues modulo 7.
(iv) If p is any prime, every number a with $1 \leqslant a \leqslant p - 1$ is relatively prime to p. This is because p cannot divide such an a, so 1 must be the only common divisor of a and p. The set of numbers $\{1, \ldots, p - 1\}$ is therefore a reduced set of residues modulo p and thus $\phi(p) = p - 1$.

Suppose now that the set $S = \{x_1, \ldots, x_r\}$ is a reduced (or complete) set of residues modulo m (so $r = \phi(m)$ in the reduced case; $r = m$ in the complete case). If a is any integer relatively prime to m we shall show that the set $T = \{ax_1, ax_2, \ldots, ax_r\}$ is also a reduced (or complete) set of residues modulo m.

Certainly no two different numbers in T are congruent, because $ax_i \equiv ax_j \pmod{m}$ would imply $x_i \equiv x_j \pmod{m}$ and this is only true if $i = j$. This observation is enough to give the result if S is a complete set of residues, for then T consists of m integers which are distinct modulo m, so T is also a complete set of residues.

If S is a reduced set of residues then T contains $\phi(m)$ integers, no two of which are congruent modulo m. Moreover, for each i with $1 \leqslant i \leqslant r, (ax_i, m) = 1$ since $(a, m) = 1$ and $(x_i, m) = 1$.

Thus the elements of T must be congruent, in some order to all the relatively prime remainders, and this means that T is a reduced set of residues.

We can now establish a classical result about $\phi(m)$ which has important consequences for some types of polynomial congruences.

THE EULER-FERMAT THEOREM *If m is a given modulus and a is a number relatively prime to m, then*

$$a^{\phi(m)} \equiv 1 \pmod{m}. \tag{2.13}$$

Proof The set R of relatively prime remainders $r_1, \ldots, r_{\phi(m)}$ is a reduced set of residues modulo m. Since $(a, m) = 1$, the set of multiples $ar_1, \ldots, ar_{\phi(m)}$ is also a reduced set of residues, and so the multiples must be congruent in some order, to $r_1, \ldots, r_{\phi(m)}$. We can express this by saying that for each $i, 1 \leqslant i \leqslant \phi(m)$, we have

$$ar_i \equiv r_j \text{ for some } j \, (1 \leqslant j \leqslant \phi(m)).$$

There are $\phi(m)$ congruences of the above type, so multiplying them all together (as we may by (2.4)) we obtain

$$ar_1 \cdot ar_2 \ldots ar_{\phi(m)} \equiv r_1 \cdot r_2 \ldots r_{\phi(m)} \pmod{m},$$

or $\qquad a^{\phi(m)} \cdot (r_1 \ldots r_{\phi(m)}) \equiv r_1 \ldots r_{\phi(m)} \pmod{m}.$

Each remainder r_i is relatively prime to m, so the product $\prod\limits_{i=1}^{\phi(m)} r_i$ must be relatively prime to m. On cancelling this product from both sides of the last congruence above, we then have $a^{\phi(m)} \equiv 1 \pmod{m}$ as required.

We have already seen that $\phi(p) = p - 1$ when p is a prime, so we have the following corollary in the case of a prime modulus.

COROLLARY: FERMAT'S LITTLE THEOREM *If p is a prime number and a is relatively prime to p (that is $p \nmid a$) then*

$$a^{p-1} \equiv 1 \pmod{p}. \tag{2.14}$$

Examples

(i) $\phi(12) = 4$, and for the reduced set of residues 1, 5, 7, 11 we have $1^4 \equiv 5^4 \equiv 7^4 \equiv 11^4 \equiv 1 \pmod{12}$.

(ii) $\phi(18) = 6$ since the numbers 1, 5, 7, 11, 13, 17 form a reduced set of residues modulo 18. The sixth powers of each of these numbers are therefore congruent to 1 modulo 18.

(iii) 6 is relatively prime to 11, so from (2.14), $6^{10} \equiv 1 \pmod{11}$.

If $(a, m) = 1$ we now know that there is at least one natural number x (namely $x = \phi(m)$) for which

$$a^x \equiv 1 \pmod{m}. \qquad (2.15)$$

Indeed, any positive multiple of $\phi(m)$ will be a solution of (2.15) since if h is a natural number then

$$a^{h\phi(m)} = (a\phi^{(m)})^h \equiv 1^h \equiv 1 \pmod{m}.$$

However, for certain choices of a, there may be values of x smaller than $\phi(m)$ which also satisfy (2.15). For example, when $a = 5$, $m = 12$, we have $\phi(12) = 4$ and so $5^4 \equiv 1 \pmod{12}$, but actually here $5^2 = 25 \equiv 1 \pmod{12}$. In general **if $(a, m) = 1$ and k is the smallest natural number which is a solution of (2.15), then k is called the order of a modulo m.** Every natural number n such that $a^n \equiv 1 \pmod{m}$ must be a multiple of the order k of a. For if such an n were not divisible by k we should have $n = qk + r$ with $0 < r < k$. This gives $1 \equiv a^n = (a^k)^q \cdot a^r \pmod{m}$ whence $a^r \equiv 1 \pmod{m}$; but this is not possible since r is a natural number smaller than k and k is the smallest solution of (2.15). In particular, we see that the order of a number relatively prime to m must be a divisor of $\phi(m)$.

Examples

(i) 2 is of order 3 modulo 7, because 2^3 is the first power of 2 which is congruent to 1 modulo 7.

(ii) 5 is of order 6 $(= \phi(18))$ modulo 18.

(iii) Any number relatively prime to 11 must be of order 1, 2, 5 or 10 since these are the only divisors of $\phi(11) = 10$. For the numbers a in the reduced set of residues given by $1 \leqslant a \leqslant 10$ we find that 1 is of order 1; 10 is of order 2; 3, 4, 5, 9 are of order 5; and 2, 6, 7, 8 are of order 10.

The last example above shows that for some moduli m there are numbers of each allowable order; in particular, numbers which have the largest possible order $\phi(m)$. This is not always the case. When $m = 12$ there is no number whose order is $\phi(12)$, because each number relatively prime to 12 has order 1 or 2. If, for a given m, there is a number a whose order modulo m is $\phi(m)$, we say that a is a **primitive root** modulo m. It is possible to prove that the only moduli which have primitive roots are 1, 2, 4, powers of a single odd prime, and twice such powers.* We shall see in Chapter 3 that when primitive roots exist they greatly facilitate the search for solutions of certain types of polynomial congruences.

Examples
(i) 10 is twice an odd prime and so has primitive roots. We have $\phi(10) = 4$ and 3 is of order 4 modulo 10.
(ii) None of the numbers $15 = 3 \cdot 5$, $20 = 2^2 \cdot 5$, $21 = 3 \cdot 7$, has primitive roots.
(iii) $27 = 3^3$ has primitive roots. Here we find that $\phi(27) = 18$ and 2 is of order 18 modulo 27.

Fermat's Little Theorem itself has an interesting consequence for polynomial congruences with respect to a prime modulus p. First of all note that each natural number n is congruent, modulo $p - 1$, to one of the numbers in the complete set of residues $\{1, \dots, p - 1\}$. In other words we can write $n = q(p - 1) + r$, where q and r are integers and $1 \leqslant r \leqslant p - 1$. Now if x is a number not divisible by p we see from Fermat's theorem that

$$x^n = (x^{p-1})^q \cdot x^r \equiv 1^q \cdot x^r \equiv x^r \pmod{p}.$$

If $p \mid x$, then $x \equiv 0 \pmod{p}$, whence $x^n \equiv 0 \pmod{p}$ and $x^r \equiv 0 \pmod{p}$, so again in this case $x^n \equiv x^r \pmod{p}$. This last congruence is therefore an identity which is true irrespective of the value of x. Thus if x^n occurs in a congruence modulo p, we may replace it by x^r without changing the

* See Appendix.

solutions of the congruence. By similarly reducing the degree of each term in a polynomial congruence we see that **any polynomial in x is identically congruent modulo p to a polynomial of degree less than p.**

Example 1   For any prime p we have the identity $x^p \equiv x \pmod{p}$. This is the same as saying that each of the p integers in any complete set of residues is a solution of $x^p - x \equiv 0 \pmod{p}$.

Example 2   Reduce the degree of the congruence $x^5 + x^4 - 7 \equiv 0$ (mod 3). Here $p - 1 = 2$ and $5 \equiv 1 \pmod{p - 1}$, $4 \equiv 2 \pmod{p - 1}$. Thus $x^5 + x^4 - 7 \equiv x^2 + x + 2 \pmod 3$ and it is easy to check that $x^2 + x + 2 \equiv 0 \pmod 3$ has no solutions.

Example 3   Simplify the congruence $f(x) \equiv 0 \pmod 7$ where $f(x) = 9x^{15} - 6x^{11} + x^2 + 23$. Here we have $x^{15} \equiv x^3 \pmod 7$, and $x^{11} \equiv x^5 \pmod 7$, so $f(x) \equiv 9x^3 - 6x^5 + x^2 + 2 \equiv x^5 + 2x^3 + x^2 + 2 \pmod 7$. Since 7 divides $x^5 + 2x^3 + x^2 + 2 = (x^2 + 2)(x^3 + 1)$ only when $7 \mid x^2 + 2$ or $7 \mid x^3 + 1$, we see that $x^2 + 2 \equiv 0 \pmod 7$ or $x^3 + 1 \equiv 0 \pmod 7$. The first of these last two congruences has no solutions; the second has the three solutions (in any complete set of residues) $x \equiv 3, 5$ and $6 \pmod 7$. These are therefore the solutions of the original congruence $f(x) \equiv 0 \pmod 7$.

If we know the prime decomposition of a natural number n, then we can calculate $\phi(n)$ directly without first finding a reduced set of residues. This is possible because as we shall see below, **ϕ is a multiplicative function.** That is,*

$$(a, b) = 1 \text{ implies } \phi(ab) = \phi(a) \cdot \phi(b). \qquad (2.16)$$

So if $n = p_1^{\alpha_1} \ldots p_r^{\alpha_r}$, where $p_1, \ldots, p_r$ are distinct prime numbers, then (2.16) gives

$$\phi(n) = \phi(p_1^{\alpha_1}) \ldots \phi(p_r^{\alpha_r}).$$

Now for any prime power p^α, $\phi(p^\alpha)$ is, by definition, the number of natural numbers which lie between 1 and p^α. and which are relatively prime to p^α. That is, the number of integers in the given range which are not multiples of p.

---

* Compare this with equation (1.25).

Since there are $\dfrac{p^{\alpha}}{p} = p^{\alpha-1}$ multiples of p between 1 and p^{α}, we must have

$$\phi(p^{\alpha}) = p^{\alpha} - p^{\alpha-1} = p^{\alpha}\left(1 - \frac{1}{p}\right). \qquad (2.17)$$

Using (2.17) in our previous expression for $\phi(n)$, we see that

$$\phi(n) = p_1^{\alpha_1}\left(1 - \frac{1}{p_1}\right)p_2^{\alpha_2}\left(1 - \frac{1}{p_2}\right)\ldots p_r^{\alpha_r}\left(1 - \frac{1}{p_r}\right)$$

$$= p_1^{\alpha_1}\ldots p_r^{\alpha_r}\left(1 - \frac{1}{p_1}\right)\ldots\left(1 - \frac{1}{p_r}\right),$$

or $\qquad \phi(n) = n\displaystyle\prod_{i=1}^{r}\left(1 - \frac{1}{p_i}\right). \qquad (2.18)$

Example Multiplying out the product in (2.18) gives

$$\phi(n) = n\left[1 - \frac{1}{p_1} - \ldots - \frac{1}{p_r} + \frac{1}{p_1 p_2} + \ldots + \frac{1}{p_{r-1}p_r} - \frac{1}{p_1 p_2 p_3} - \ldots\right]$$

$$= n\left[1 + \sum_{s=1}^{r}\ \sum_{p_{i_1},\ldots,\,p_{i_s}}\frac{(-1)^s}{(p_{i_1}\cdots p_{i_s})}\right],$$

where, for each s, the primes $p_{i_1}, \ldots, p_{i_s}$ range over every choice of s distinct primes from among $p_1, \ldots, p_r$. This can be written as $n\displaystyle\sum_{s=0}^{r}\ \sum_{p_{i_1},\ldots,\,p_{i_s}}\frac{\mu(p_{i_1}\cdots p_{i_s})}{(p_{i_1}\cdots p_{i_s})}$, or as $n\displaystyle\sum_{h\mid n}\frac{\mu(h)}{h}$, since $\mu(h) = 0$ if h is not a product of distinct primes. Therefore $\phi(n) = \displaystyle\sum_{h\mid n}\mu(h)\cdot\frac{n}{h}$. This is a particular case of the Möbius formula $g(n) = \displaystyle\sum_{h\mid n}\mu(h)f\left(\frac{n}{h}\right)$, where $g(n) = \phi(n)$ and $f(n) = n$. We can thus apply the inverse formula $f(n) = \displaystyle\sum_{h\mid n}g(h)$ and conclude that

$$n = \sum_{h\mid n}\phi(h). \qquad (2.19)$$

We have derived the formula (2.18) for $\phi(n)$ from the fact that ϕ is a multiplicative function. In order to see that ϕ is multiplicative we shall suppose that a and b are two relatively prime numbers and that $\{x_1, \ldots, x_{\phi(a)}\}$, $\{y_1, \ldots, y_{\phi(b)}\}$ are reduced sets of residues modulo a and b respectively. We shall combine these sets of residues to produce a reduced set of residues modulo ab, which will contain $\phi(a) . \phi(b)$ integers. This will immediately give the desired result that $\phi(ab) = \phi(a) . \phi(b)$.

The numbers forming our reduced set of residues modulo ab will actually be all those of the form $bx_i + ay_j$, where x_i is any one of $x_1, \ldots, x_{\phi(a)}$ and y_j is one of $y_1, \ldots, y_{\phi(b)}$. There are indeed $\phi(a) . \phi(b)$ such numbers, because corresponding to each of the $\phi(a)$ choices for x_i there are $\phi(b)$ choices of y_j, and so there are $\phi(a) . \phi(b)$ different possibilities for the expression $bx_i + ay_j$.

No two of the numbers $bx_i + ay_j$ are congruent modulo ab. If we had, say, $bx_i + ay_j \equiv bx_i' + ay_j' \pmod{ab}$, then we would certainly have $bx_i + ay_j \equiv bx_i' + ay_j' \pmod{a}$. This gives $bx_i \equiv bx_i' \pmod{a}$, whence $x_i \equiv x_i' \pmod{a}$ since $(b, a) = 1$. This actually means $x_i = x_i'$, because no two different members of the set $\{x_1, \ldots, x_{\phi(a)}\}$ are congruent modulo a. Similarly we would have $y_j \equiv y_j' \pmod{b}$ and so $y_j = y_j'$. Thus two of the numbers $bx_i + ay_j$ cannot be congruent modulo ab if they are formed from different x's or different y's.

Also, each of the expressions $bx_i + ay_j$ must be relatively prime to ab. For if we suppose, on the contrary, that for some particular x_i and y_j, ab and $bx_i + ay_j$ had a common divisor d greater than one, then any prime divisor p of d would divide both ab and $bx_i + ay_j$. So we would have either p divides both a and $bx_i + ay_j$ or p divides both b and $bx_i + ay_j$. The first alternative implies $p \mid bx_i$, but this is impossible with p a divisor of a since both b and x_i are relatively prime to a. Similarly, the second alternative is impossible and therefore we must have $(ab, bx_i + ay_j) = 1$.

We have now shown that, on division by ab, the numbers

$bx_i + ay_j$ leave some of the relatively prime remainders $r_1, \ldots r_{\phi(ab)}$. In order to see that they include all the relatively prime remainders we must show that each number m, relatively prime to ab, is congruent (mod ab) to one of them. First we note that if $(m, ab) = 1$ then $(m, a) = 1$ and $(m, b) = 1$. Next, because $(a, b) = 1$, the numbers $bx_1, \ldots, bx_{\phi(a)}$ form a reduced set of residues modulo a so there must be one of them, say bx_i, such that $bx_i \equiv m \pmod{a}$. Similarly there must be a y_j such that $ay_j \equiv m \pmod{b}$. So for this particular i and j, $bx_i + ay_j$ is congruent to m modulo a and modulo b and thus modulo ab. For, $bx_i + ay_j - m$ is divisible by a and by b, and thus, since $(a, b) = 1$, by ab. Therefore our $\phi(a) \cdot \phi(b)$ linear combinations do form a reduced set of residues modulo ab. As noted earlier this shows $\phi(ab) = \phi(a) \cdot \phi(b)$ as required.

Example 1 $21 = 3 \cdot 7$, so from (2.18) we have $\phi(21) = 21(1 - \frac{1}{3})(1 - \frac{1}{7}) = 12$. Similarly since $12 = 2^2 \cdot 3$ and $27 = 3^3$ we have $\phi(12) = 12(1 - \frac{1}{2})(1 - \frac{1}{3}) = 4$ and $\phi(27) = 27(1 - \frac{1}{3}) = 18$.

Example 2   Since $561 = 3 \cdot 11 \cdot 17$ we have

$$\phi(561) = 3 \cdot 11 \cdot 17 \left(\tfrac{2}{3}\right)\left(\tfrac{10}{11}\right)\left(\tfrac{16}{17}\right) = 2 \cdot 10 \cdot 16 = 320,$$

so every number a relatively prime to 561 satisfies $a^{320} \equiv 1 \pmod{561}$. However, there is no number whose order modulo 561 is 320, and we can actually show that the order of each number relatively prime to 561 is a divisor of 80. This is because if $(a, 561) = 1$ (that is, $3 \nmid a$, $11 \nmid a$ and $17 \nmid a$) then from Fermat's Little Theorem we see that $a^{80} = (a^2)^{40} \equiv 1 \pmod{3}$, $a^{80} = (a^{10})^8 \equiv 1 \pmod{11}$ and $a^{80} = (a^{16})^5 \equiv 1 \pmod{17}$. Thus $a^{80} - 1$ must be divisible by the product of 3, 11 and 17, which gives

$$a^{80} \equiv 1 \pmod{561}.$$

An interesting consequence is that (because 80 happens to divide 560) we have

$$a^{560} \equiv 1 \pmod{561} \text{ if } (a, 561) = 1.$$

This last congruence is now reminiscent of Fermat's theorem, which says that if $(a, n) = 1$ then the congruence

$$a^{n-1} \equiv 1 \pmod{n} \tag{2.20}$$

is always true if n is a prime number. We now know that (2.20) can also hold

43

CONGRUENCES AND EQUATIONS

sometimes if n is composite. The only composite numbers n, less than 2000, which satisfy (2.20) are 561, $1105 = 5 . 13 . 17$ and $1729 = 7 . 13 . 19$.

Example 3   Construct a reduced set of residues modulo $12 = 3 . 4$ from reduced sets of residues modulo 3 and 4. The sets $\{1, 2\}$ and $\{1, 3\}$ are reduced sets of residues modulo 3 and 4 respectively. So we know that all the numbers of the form $4x + 3y$, where x is either 1 or 2 and y is either 1 or 3, will constitute a reduced set of residues modulo 12. These numbers are $4 + 3 = 7$, $4 . 2 + 3 = 11$, $4 + 3 . 3 = 13$, and $4 . 2 + 3 . 3 = 17$.

5. Simplifying the modulus of a congruence

Suppose m is a given modulus and a and b are numbers such that

$$a \equiv b \pmod{m}. \tag{2.21}$$

If $m = p_1^{\alpha_1} \ldots p_r^{\alpha_r}$, where $p_1, \ldots, p_r$ are distinct prime numbers, then from $m \mid a - b$ it is apparent that $p_i^{\alpha_i} \mid a - b$ for each of the prime powers $p_i^{\alpha_i}$ occurring in the factorization of m. So the single congruence (2.21) implies that each of the following r congruences must hold:

$$\begin{aligned} a &\equiv b \pmod{p_1^{\alpha_1}} \\ &\cdots \\ a &\equiv b \pmod{p_r^{\alpha_r}}. \end{aligned} \tag{2.22}$$

Conversely, if we are given the r congruences in (2.22), then $a - b$ must be divisible by each of the prime powers $p_i^{\alpha_i}$ $(1 \leqslant i \leqslant r)$. Since the distinct prime powers are relatively prime, it follows that $a - b$ must be divisible by their product, namely m. So the congruence (2.21) must hold. In other words the single congruence (2.21) and the set of congruences (2.22) are equivalent and we may replace one of them by the other whenever we wish.

Example $14 \equiv 74 \pmod{60}$ implies and is implied by the three congruences $14 \equiv 74 \pmod 4$, $14 \equiv 74 \pmod 3$ and $14 \equiv 74 \pmod 5$.

Consider now the polynomial congruence

$$f(x) \equiv 0 \pmod{m}. \tag{2.23}$$

44

Using the above reasoning which relates (2.21) to (2.22), we see that finding an integer x satisfying (2.23) is exactly the same as finding a single integer x which simultaneously satisfies the following congruences

$$f(x) \equiv 0 \pmod{p_1^{\alpha_1}}$$
$$\cdots \qquad\qquad (2.24)$$
$$f(x) \equiv 0 \pmod{p_r^{\alpha_r}}.$$

In practice, it is usually simpler to solve the congruences (2.24) separately. That is, to find r integers $a_1, \ldots, a_r$ such that $f(a_1) \equiv 0 \pmod{p_1^{\alpha_1}}$, $f(a_2) \equiv 0 \pmod{p_2^{\alpha_2}} \ldots$, and $f(a_r) \equiv 0 \pmod{p_r^{\alpha_r}}$. We shall see that these separate solutions can always be used to construct a common solution.

Example 1   Consider the congruence $x^2 \equiv 7 \pmod{60}$. Since $60 = 2^2 . 3 . 5$ we must find an integer x which simultaneously satisfies the three congruences $x^2 \equiv 7 \pmod 4$, $x^2 \equiv 7 \pmod 3$ and $x^2 \equiv 7 \pmod 5$. But on testing the numbers in any complete set of residues modulo 4 we see that $x^2 \equiv 7 \pmod 4$ has no solutions (and neither has $x^2 \equiv 7 \pmod 5$). There are therefore no solutions to the original congruence $x^2 \equiv 7 \pmod{60}$.

Example 2   Solve the congruence $f(x) = x^3 + 9x^2 + x - 5 \equiv 0 \pmod{105}$. Since $105 = 3 . 5 . 7$ we must solve the three congruences $f(x) \equiv 0 \pmod 3$, $f(x) \equiv 0 \pmod 5$ and $f(x) \equiv 0 \pmod 7$. On examining these congruences separately we find that $f(1) \equiv 0 \pmod 3$, $f(0) \equiv 0 \pmod 5$, and $f(5) \equiv 0 \pmod 7$ but it is not immediately apparent that they have a common solution. In this case $x = 40$ satisfies all three congruences and is thus a solution of the original congruence modulo 105.

Any polynomial congruence behaves like one or other of the above examples. For certainly if one or more of the congruences (2.24) has no solution, then the congruence (2.23) has no solution. On the other hand, **if all the congruences (2.24) can be solved separately then they can always be solved simultaneously** and (2.23) will have a solution. This is a simple consequence of the following famous result about simultaneous congruences.

THE CHINESE REMAINDER THEOREM   *Suppose that $m_1, \ldots, m_r$ are natural numbers every two of which are relatively prime. Then, in any complete set of residues modulo the product*

$m_1 \ldots m_r$, *there is exactly one integer which simultaneously satisfies the congruences*

$$x \equiv a_1 \ (\text{mod } m_1)$$
$$\ldots \qquad\qquad (2.25)$$
$$x \equiv a_r \ (\text{mod } m_r).$$

Proof We shall denote the product $m_1 \ldots m_r$ by M and for each i $(1 \leqslant i \leqslant r)$ we shall denote the quotient $M/m_i = m_1 \ldots m_{i-1}m_{i+1} \ldots m_r$ by M_i. Since M_i only involves the moduli relatively prime to m_i, we have $(M_i, m_i) = 1$, so there will be an integer v_i such that $M_i v_i \equiv 1 \ (\text{mod } m_i)$. Now consider the integer x_0 given by

$$x_0 = a_1 M_1 v_1 + a_2 M_2 v_2 + \ldots + a_r M_r v_r.$$

Because m_1 divides each of $M_2, \ldots, M_r$, we have $x_0 \equiv a_1 M_1 v_1 \equiv a_1 \ (\text{mod } m_1)$. Similarly $x_0 \equiv a_i M_i v_i \equiv a_i \ (\text{mod } m_i)$ for each modulus m_i, so x_0 is a common solution of the congruences (2.25). Any integer which differs from x_0 by a multiple of M will also be a solution of the congruences (2.25), since any multiple of M will be a multiple of each modulus m_i.

In any complete set of residues modulo M the integer congruent to $x_0 \ (\text{mod } M)$ will be the only integer satisfying all the congruences (2.25). This is because if x_1 is any solution of (2.25) we shall have $x_1 \equiv a_i \equiv x_0 \ (\text{mod } m_i)$ for each modulus m_i, which gives $m_i \mid x_1 - x_0$. Therefore $x_1 - x_0$ is divisible by the product M, or $x_1 \equiv x_0 \ (\text{mod } M)$ as required.

Example Find the smallest natural number which leaves remainder 3 on division by 4 and remainder 5 on division by 7. Since 4 and 7 are relatively prime, the Chinese Remainder Theorem assures us that there is exactly one solution between 1 and 28 inclusive. Instead of using the general method of the theorem, we can find the solution by arguing that the only numbers between 1 and 28 which are congruent to 5 (mod 7) are 5, 12, 19 and 26. Of these, 19 is congruent to 3 modulo 4, so 19 is the smallest natural number which satisfies both conditions.

For any polynomial $f(x)$, the Chinese Remainder Theorem

enables us to construct a simultaneous solution of the congruences (2.24) from separate solutions $a_1, \ldots, a_r$. Because, if $a_1, \ldots, a_r$ are integers such that $f(a_1) \equiv 0 \pmod{p_1^{\alpha_1}}, \ldots,$ $f(a_r) \equiv 0 \pmod{p_r^{\alpha_r}}$, then we now know that we can choose an integer a which satisfies

$$a \equiv a_1 \pmod{p_1^{\alpha_1}}, \ldots, a \equiv a_r \pmod{p_r^{\alpha_r}}.$$

This implies that

$$f(a) \equiv f(a_1) \equiv 0 \pmod{p_1^{\alpha_1}}, \ldots, f(a) \equiv f(a_r) \equiv 0 \pmod{p_r^{\alpha_r}},$$

so $x = a$ is a solution of all the congruences (2.24) and is therefore a solution of the congruence (2.23).

We can now solve any polynomial congruence, provided only that we can solve congruences to prime power moduli, such as

$$f(x) \equiv 0 \pmod{p^\alpha} \tag{2.26}$$

where p is a prime and α is a natural number. In order to be able to solve (2.26) we certainly have to be able to solve the congruence

$$f(x) \equiv 0 \pmod{p} \tag{2.27}$$

since any integer satisfying (2.26) automatically satisfies (2.27). When $\alpha \geqslant 2$ it may sometimes happen that even if (2.27) is soluble, (2.26) is not soluble. For example, $x^2 \equiv 5 \pmod{2}$ is soluble (with $x = 1$) but $x^2 \equiv 5 \pmod{8}$ has no solutions. It is possible to say when the solubility of (2.27) implies the solubility of (2.26), but we shall not attempt here to describe the conditions which must be satisfied. In Chapter 3 we shall concentrate on polynomial congruences (and especially quadratic congruences) modulo a prime. We shall see later how a knowledge of the theory of quadratic congruences to a prime modulus leads to several interesting results about sums of squares, such as the fact that **every natural number can be expressed as the sum of at most four squares of natural numbers.**

EXERCISES

1. Suppose the number N can be written in the decimal notation as $c_n c_{n-1} \ldots c_0$. Use the fact that $10 \equiv -1 \pmod{11}$ to deduce a simple criterion for N to be divisible by 11.

2. The number $2^{127} - 1$ is known to be prime Find the last digit of its decimal representation.

3. Find the solutions (if any) of the following congruences $f(x) \equiv 0 \pmod{m}$. In each case, what can be said about the solutions of the corresponding equation $f(x) = 0$?

   (i) $f(x) = x^2 - 10x - 119, m = 8$;
   (ii) $f(x) = 14x^3 - 7x^2 + 3x + 1, m = 2$;
   (iii) $f(x) = x^4 - 31x^3 + 252x^2 - 251x + 221, m = 13$;
   (iv) $f(x) = (2x + 1)(3x + 1), m = 5$.

4. Find the solutions (if any) of the following linear congruences:

   (i) $2x \equiv 1 \pmod{9}$;   (ii) $3x \equiv 7 \pmod{12}$;
   (iii) $21x \equiv 15 \pmod{18}$;   (iv) $6x \equiv 3 \pmod{7}$.

5. Construct a reduced set of residues modulo 30 from reduced sets of residues modulo 5 and 6.

6. It has been conjectured that for every natural number n there is another number $m \neq n$ such that $\phi(n) = \phi(m)$. It is not known whether this is true or false in general, but prove that it is true if n is an odd prime. Is the conjecture true if n is any odd number?

7. Prove that 3 is a primitive root modulo 50.

8. Let the integer α be at least 3. Prove by induction that 3 is of order $2^{\alpha-2}$ modulo 2^α and also that $3^{2^{\alpha-2}} = 1 + 2^\alpha t_\alpha$ for some odd integer t_α.

9. Solve the following simultaneous congruences:

   (i) $2x \equiv 1 \pmod{9}$;   (ii) $x \equiv 1 \pmod{3}$;   (iii) $2x \equiv 6 \pmod{8}$.
       $5x \equiv 3 \pmod{4}$ $x \equiv 3 \pmod{5}$ $x \equiv 1 \pmod{5}$
                                      $x \equiv 5 \pmod{7}$

10. Find the solutions of $2x^9 + 2x^8 + x^3 + 1 \equiv 0 \pmod{5}$.

11. Solve the congruence $4x^4 + 4x^3 + 6x^2 + 21x + 7 \equiv 0 \pmod{84}$.

CHAPTER THREE

Non-linear Congruences

1. Introduction

In Chapter 2 we saw that it is always possible to tell whether or not a linear congruence is soluble without actually searching for solutions. In this chapter we shall investigate polynomial congruences of degree two or more (non-linear congruences) with prime moduli, and ask similarly if we can say whether or not a given non-linear congruence is soluble without first having to look for possible solutions. We shall see that there is a complete theory for congruences of degree two, and this will enable us to ascertain fairly easily if any given quadratic congruence is soluble. There is no theory which covers all congruences of degree three or more so we shall only examine one general type of higher degree congruence.

2. General results

There is one simple and often very useful result which applies to any polynomial congruence with a prime modulus. Namely, if $f(x)$ is a polynomial of degree n such that the prime p does not divide the coefficient of x^n in $f(x)$, then **the congruence $f(x) \equiv 0 \pmod{p}$ has no more than n distinct solutions modulo p.** We deduce this result by using induction on the degree n of $f(x)$. So, supposing always that $f(x)$ satisfies the necessary hypothesis, assume first that $n = 1$ and thus that $f(x)$ is linear.

Then p is relatively prime to the coefficient of x in $f(x)$, and we have already seen in our discussion of linear congruences that in this case there is exactly one solution. Assume, then, that $n \geqslant 2$ and that the result is true for all congruences of degree less than n. If $f(x) \equiv 0 \pmod{p}$ has no solutions then the result is certainly true. If there is at least one solution, $x = s$ say, then we can divide $f(x)$ by $x - s$ to obtain

$$f(x) = (x - s)g(x) + c$$

where $g(x)$ is a polynomial of degree $n - 1$ with integer coefficients and the remainder c is an integer. The leading coefficient of $g(x)$—being the same as that of $f(x)$—is not divisible by p, so, by the inductive hypothesis, $g(x) \equiv 0 \pmod{p}$ has at most $n - 1$ solutions in a complete set of residues. Also $f(s) = c \equiv 0 \pmod{p}$, which means that $f(x) \equiv 0 \pmod{p}$ becomes $(x - s)g(x) \equiv 0 \pmod{p}$ whence, since p is prime, we must have either $x - s \equiv 0 \pmod{p}$ or $g(x) \equiv 0 \pmod{p}$. Therefore, apart possibly from s, all the solutions of the original congruence must satisfy $g(x) \equiv 0 \pmod{p}$, so their total number is at most $1 + (n - 1) = n$ as required.

Example 1   The congruence $x^p - x \equiv 0 \pmod{p}$ can have no more than p distinct solutions. In this case we already know from Fermat's theorem that the congruence has its maximum allowable number of solutions. Indeed, applying the above process of factoring out each root in turn we see that

$$x^p - x \equiv x(x - 1)(x - 2) \ldots (x - \overline{p - 1}) \pmod{p}. \qquad (3.1)$$

Example 2   The congruence $x^2 \equiv 1 \pmod{p}$ can have at most two solutions. Since $x \equiv 1 \pmod{p}$ and $x \equiv -1 \pmod{p}$ are each solutions and if $p > 2$ they are distinct modulo p, we see that they are the only solutions.

Example 3   The result is not immediately applicable to the congruence $f(x) \equiv 0 \pmod{7}$ if $f(x) = 7x^6 - 21x^4 + x^3 - x^2 - 4x - 6$. But $f(x) \equiv x^3 - x^2 - 4x - 6 \pmod{7}$ so the congruence has at most three solutions. Actually it only has one: $x \equiv 3 \pmod{7}$. In general if the leading coefficient of $f(x)$ is divisible by p then we are really dealing with a polynomial of lower degree than at first thought. So in trying to solve polynomial congruences of a certain fixed degree, there is no loss in always assuming that p does not divide the leading coefficient of the polynomial.

No polynomial congruence modulo p can have more than p distinct solutions since that is the number of integers in any complete set of residues. So the previous result will only be really useful when the degree of the polynomial in question is less than p. However, we already know that nothing will be lost by only considering such polynomials, since Fermat's theorem enables us to replace any polynomial congruence by an equivalent one having the same solutions and in which the degree of the polynomial is smaller than p. Fermat's theorem also enables us to dispose of the possibility of p being 2; because any congruence $f(x) \equiv 0 \pmod{2}$ can be replaced by an equivalent one in which the left hand side is either a constant or a linear polynomial, and any solutions will then be obvious. So throughout the rest of this chapter we shall assume that p is an odd prime.

3. Completing the square in a quadratic congruence

Consider a typical quadratic congruence such as

$$Ax^2 + Bx + C \equiv 0 \pmod{p} \tag{3.2}$$

where $p \nmid A$. Multiplying both sides by $4A$ and then completing the square gives

$$(2Ax + B)^2 - (B^2 - 4AC) \equiv 0 \pmod{p}$$

or $$(2Ax + B)^2 \equiv B^2 - 4AC \pmod{p}.$$

Now if we could find an integer s such that $s^2 \equiv B^2 - 4AC \pmod{p}$ then (since $p \nmid 2A$) we could certainly solve the linear congruence $2Ax + B \equiv s \pmod{p}$ and the resulting x would satisfy (3.2). Thus we can certainly say whether or not any quadratic congruence is soluble, provided only that we know when it is possible to take square roots modulo p. So instead of dealing with a congruence like (3.2) we shall consider whether or not, for a fixed integer a, the congruence

$$x^2 \equiv a \pmod{p} \tag{3.3}$$

is soluble.

4. Indices

It is often helpful to place a particular problem in a more general setting, so we shall begin by discussing the congruence

$$x^k \equiv a \pmod{p} \tag{3.4}$$

where k is any integer greater than or equal to two. If $a \equiv 0 \pmod{p}$ then (3.4) is soluble for any k, with $x \equiv 0 \pmod{p}$ being the only solution in any complete set of residues. When $a \not\equiv 0 \pmod{p}$, that is when $(a, p) = 1$, we can gain valuable insight into (3.3) and (3.4) by using primitive roots. We remarked in Chapter 2 (and prove in the Appendix) that primitive roots exist for each prime, so we shall suppose that we have found an integer g which is a primitive root for the prime p occurring in (3.4). Then g is of order $\phi(p) = p - 1$ modulo p, which means that the $p - 1$ numbers $1 = g^0$, g, g^2, ..., g^{p-2} are all incongruent. This is because $g^i \equiv g^j \pmod{p}$ for $0 \leqslant i < j \leqslant p - 2$ would imply $g^{j-i} \equiv 1 \pmod{p}$ with $0 < j - i \leqslant p - 2$, contradicting the fact that the order of g is $p - 1$. Also, each of the $p - 1$ numbers $g^0, \ldots g^{p-2}$ is relatively prime to p, because g itself is. They therefore form a reduced set of residues modulo p, so that there is exactly one of them, say g^r, for which

$$g^r \equiv a \pmod{p} \text{ with } 0 \leqslant r \leqslant p - 2. \tag{3.5}$$

We call r the **index** of a to the base g and write $r = \text{ind}_g a$. If we ever have $g^{r'} \equiv a \pmod{p}$ for $r' \neq r$ then we must have $r' \equiv r \pmod{p - 1}$. This is because $g^{r'} \equiv a \equiv g^r \pmod{p}$ is the same as $g^{r'-r} \equiv 1 \pmod{p}$ and we know that this means $r' - r$ is divisible by the order $p - 1$ of g.

Example No matter what the value of g is, we have

(i) $\text{ind}_g\, 1 = 0$ since $g^0 = 1$;

(ii) $\text{ind}_g\,(-1) = \frac{1}{2}(p-1)$. This is because $(g^{\frac{1}{2}(p-1)})^2 = g^{p-1} \equiv 1 \pmod{p}$ so we must have either* $g^{\frac{1}{2}(p-1)} \equiv 1 \pmod{p}$ or $g^{\frac{1}{2}(p-1)} \equiv -1 \pmod{p}$. The first alternative would contradict the fact that g is of order $p-1$ so $g^{\frac{1}{2}(p-1)} \equiv -1 \pmod{p}$ and this gives the stated result.

The reason for introducing indices is that they behave very much like logarithms. Just as in ordinary arithmetic we can multiply two positive numbers together by simply adding their logarithms, so we can multiply two integers modulo p by adding their indices. To make this precise, suppose that g and g' are two primitive roots modulo p and that a and b are integers. Then the following relations hold:

$$\text{ind}_g\, a \equiv \text{ind}_g\, g' \cdot \text{ind}_{g'}\, a \pmod{p-1}, \qquad (3.6)$$

$$\text{ind}_g\,(ab) \equiv \text{ind}_g\, a + \text{ind}_g\, b \pmod{p-1}, \qquad (3.7)$$

and $\qquad \text{ind}_g\, a^n \equiv n\,\text{ind}_g\, a \pmod{p-1}. \qquad (3.8)$

In order to derive (3.6) suppose that $\text{ind}_{g'}\, a = r_1, \text{ind}_g\, g' = r_2$. Then $a \equiv (g')^{r_1} \equiv (g^{r_2})^{r_1} = g^{r_1 r_2} \pmod{p}$ giving $r_1 r_2 \equiv \text{ind}_g\, a \pmod{p-1}$ as required. Again, if $\text{ind}_g\, a = r,\, \text{ind}_g\, b = s$, we have $ab \equiv g^r \cdot g^s = g^{r+s} \pmod{p}$ so (3.7) is true. Finally (3.8) follows by induction from (3.7). This is because the case $n = 1$ is certainly true; and if we suppose that (3.8) is true when $n = t$, then the case $n = t + 1$ follows (using (3.7)) from

$$\text{ind}_g\,(a^{t+1}) \equiv \text{ind}_g\, a + \text{ind}_g\,(a^t)$$

$$\equiv \text{ind}_g\, a + t\,\text{ind}_g\, a = (t + 1)\,\text{ind}_g\, a \pmod{p-1}.$$

Example Show that for a, b relatively prime to p,

$$\text{ind}_g\left(\frac{a}{b}\right) \equiv \text{ind}_g\, a - \text{ind}_g\, b \pmod{p-1}.$$

In arithmetic modulo p, $\frac{a}{b}$ means an integer y such that $a \equiv by \pmod{p}$. So from (3.7) $\text{ind}_g\, a \equiv \text{ind}_g\, b + \text{ind}_g\, y \pmod{p-1}$, which gives at once $\text{ind}_g\, y \equiv \text{ind}_g\, a - \text{ind}_g\, b \pmod{p-1}$.

* See the last Example 2.

Property (3.6) is analogous to the rule for converting ordinary logarithms from one base to another. For example, for the bases e and 10 we have

$$\log_e a = \log_e 10 \, . \, \log_{10} a.$$

Corresponding to (3.7) we have of course the familiar

$$\log ab = \log a + \log b.$$

However, for this rule to be an effective aid in performing calculations, we must possess a table of logarithms. Similarly in order to make use of (3.7) and (3.8) in performing multiplication modulo p we must know the indices of all the integers in a reduced set of residues modulo p. We shall illustrate these ideas by constructing a table of indices for the prime 13. Here 2 is of order 12 modulo 13 and we shall use it as a base for our indices. There are other primitive roots modulo 13 and it is immaterial which one we take as a base.

For the base 2 we find that

$$2^0 = 1 \qquad 2^4 \equiv 3 \qquad 2^8 \equiv 9$$
$$2^1 = 2 \qquad 2^5 \equiv 6 \qquad 2^9 \equiv 5$$
$$2^2 = 4 \qquad 2^6 \equiv 12 \qquad 2^{10} \equiv 10$$
$$2^3 = 8 \qquad 2^7 \equiv 11 \qquad 2^{11} \equiv 7$$

where all the above congruences are modulo 13. We can now construct Tables I and II (taking the integers $1, \ldots, 12$ as a reduced set of residues modulo 13). Table II is just given for convenience in reconverting from indices to residues.

TABLE I

Residue	1	2	3	4	5	6	7	8	9	10	11	12
Index	0	1	4	2	9	5	11	3	8	10	7	6

TABLE II

Index	0	1	2	3	4	5	6	7	8	9	10	11
Residue	1	2	4	8	3	6	12	11	9	5	10	7

Example 1 Use indices to multiply 3 by 7 (mod 13). From Table I, $\text{ind}_2 3 = 4$ and $\text{ind}_2 7 = 11$ so (3.7) gives

$$\text{ind}_2 (3 . 7) \equiv 4 + 11 \equiv 3 \text{ (mod 12)}.$$

Table II then implies $3 . 7 \equiv 8 \text{ (mod 13)}$.

Example 2 Solve the linear congruence $5x \equiv 9 \text{ (mod 13)}$. Here (3.7) gives $\text{ind}_2 5 + \text{ind}_2 x \equiv \text{ind}_2 9 \text{ (mod 12)}$ whence $\text{ind}_2 x \equiv \text{ind}_2 9 - \text{ind}_2 5 = 8 - 9 \equiv 11 \text{ (mod 12)}$, and we infer from Table II that $x \equiv 7 \text{ (mod 13)}$.

Example 3 Solve the congruence $6x^3 \equiv 7 \text{ (mod 13)}$. Using (3.7) and (3.8) we have $\text{ind}_2 6 + 3 \text{ind}_2 x \equiv \text{ind}_2 7 \text{ (mod 12)}$. Thus $3 \text{ind}_2 x \equiv 11 - 5 = 6$ (mod 12) so that $\text{ind}_2 x \equiv 2 \text{ (mod 4)}$ from (2.7). Since $\text{ind}_2 x$ lies between 0 and 11 inclusive, it can be either 2, 6 or 10, so Table II shows that there are the three solutions $x \equiv 4$, $x \equiv 12$ and $x \equiv 10 \text{ (mod 13)}$.

In the case of (3.4) we shall assume that a primitive root g modulo p has been chosen and that $r = \text{ind}_g a$. Then taking indices of both sides, $x^k \equiv a \text{ (mod } p)$ becomes

$$k \, \text{ind}_g x \equiv r \text{ (mod } p - 1). \tag{3.9}$$

If $(k, p - 1) = 1$, then for every r, (3.9) is uniquely soluble for $\text{ind}_g x$ so that (3.4) is soluble for every value of a. If $(k, p - 1) = d > 1$, (3.9) is only soluble when r is one of the $\dfrac{p - 1}{d}$ numbers in the range $0 \leqslant r \leqslant p - 2$ which are divisible by d. When (3.9) (and so (3.4)) is soluble we say that a is a **kth power residue modulo p** and when (3.4) is insoluble we say that a is a **kth power non-residue modulo p**.

5. Quadratic residues and the calculation of the Legendre symbol

From now on we shall restrict our attention to $k = 2$, and so to (3.3). When (3.3) is soluble with $a \not\equiv 0 \pmod{p}$ we shall say that a is a **quadratic residue modulo p,** and when (3.3) is insoluble we shall say that a is a **quadratic non-residue modulo p.** In mathematics, as in other disciplines, a concept is often easier to deal with once it has been given a name. In this case (3.9) assumes the form

$$2 \operatorname{ind}_g x \equiv r \pmod{p - 1}.$$

Here $2 \mid p - 1$ since p is odd, so we see that a is only a quadratic residue if it is one of the $\frac{1}{2}(p - 1)$ numbers (in any given reduced set of residues modulo p) whose index is even. When a is one of the $\frac{1}{2}(p - 1)$ numbers whose index is odd it is a quadratic non-residue. In particular, in any reduced set of residues there are the same number of quadratic residues as of quadratic non-residues.

Example 1 -1 is a quadratic residue modulo p if $p \equiv 1 \pmod 4$ and a non-residue if $p = -1 \pmod 4$. This is because the index of -1 is always $\frac{1}{2}(p - 1)$, which is even if $p \equiv 1 \pmod 4$, and odd if $p \equiv -1 \pmod 4$.

Example 2   In the reduced set of residues $\{1, 2, \ldots, 6\} \pmod 7$ the numbers 1, 2, 4 are quadratic residues, since $1^2 = 1$, $2^2 = 4$ and $3^2 \equiv 2 \pmod 7$. The numbers 3, 5, 6 are quadratic non-residues.

Example 3   If a is the square of some natural number k it is a quadratic residue for any prime p with $p \nmid a$ because (3.3) is then always soluble with $x = k$.

Mathematical advances are often made by being able to translate a concept described in words into a convenient symbol which can be manipulated according to certain rules. In our present problem Legendre's* symbol $\left(\dfrac{a}{p} \right)$ neatly

* 1752–1833.

specifies the property of being a quadratic residue or non-residue of a given prime. It is defined for any integer a and any odd prime p as follows:

$$\left(\frac{a}{p}\right) = \begin{cases} +1 \text{ if } a \text{ is a quadratic residue modulo } p \\ -1 \text{ if } a \text{ is a quadratic non-residue modulo } p \quad (3.10) \\ 0 \text{ if } a \equiv 0 \pmod{p}. \end{cases}$$

Example 1 For any odd prime p we have

$$\sum_{n=1}^{p-1} \left(\frac{n}{p}\right) = 0.$$

This is a translation into symbols of the fact that in any reduced set of residues there are equal numbers of quadratic residues and quadratic non-residues. Because the set $\{1, \ldots, p-1\}$ is a reduced set of residues, the above sum contains equal numbers of $+1$'s and -1's giving a total of zero.

Example 2 If $a \equiv b \pmod{p}$ then

$$\left(\frac{a}{p}\right) = \left(\frac{b}{p}\right). \quad (3.11)$$

This is simply because the solubility, or insolubility, of (3.3) is unchanged by replacing a by b.

Our original problem of determining when (3.3) is soluble has now become the problem of evaluating $\left(\frac{a}{p}\right)$ for any given a and p. As a first step we deduce

EULER'S CRITERION *For any integer a and odd prime number p,*

$$\left(\frac{a}{p}\right) \equiv a^{\frac{1}{2}(p-1)} \pmod{p}. \quad (3.12)$$

Equation (3.12) is certainly true if $p \mid a$, for then both sides are congruent to zero modulo p. It is also true if a is a quadratic residue; for if, say, $x_1^2 \equiv a \pmod{p}$, then $a^{\frac{1}{2}(p-1)} \equiv (x_1^2)^{\frac{1}{2}(p-1)} = x_1^{p-1} \equiv 1 = \left(\frac{a}{p}\right) \pmod{p}$. This also shows that in any reduced set of residues modulo p the quadratic residues are the only solutions of the congruence $x^{\frac{1}{2}(p-1)} \equiv 1 \pmod{p}$. Because, being of degree $\frac{1}{2}(p-1)$, the congruence can have no more than $\frac{1}{2}(p-1)$ solutions, and we have just seen that each of the $\frac{1}{2}(p-1)$ quadratic residues is a solution. Finally if b is a quadratic non-residue, $b^{p-1} \equiv 1 \pmod{p}$ implies that either $b^{\frac{1}{2}(p-1)} \equiv 1 \pmod{p}$ or $b^{\frac{1}{2}(p-1)} \equiv -1 \pmod{p}$. We cannot have $b^{\frac{1}{2}(p-1)} \equiv 1 \pmod{p}$ as this would only be true for a quadratic residue b. Hence $b^{\frac{1}{2}(p-1)} \equiv -1 \pmod{p}$ holds, in agreement with (3.12).

Example We can use Euler's criterion, if we wish, to find when -1 is a quadratic residue without using indices. This is because it gives $\left(\frac{-1}{p}\right) \equiv (-1)^{\frac{1}{2}(p-1)} \pmod{p}$, so actually $\left(\frac{-1}{p}\right) = (-1)^{\frac{1}{2}(p-1)}$ (since if one side were $+1$ and the other -1, they would be incongruent). The result now follows, for $(-1)^{\frac{1}{2}(p-1)}$ is equal to $+1$ if $p \equiv 1 \pmod{4}$, and equal to -1 if $p \equiv -1 \pmod{4}$.

Euler's criterion does not, by itself, provide a very practical means of calculating $\left(\frac{a}{p}\right)$ if a and p are at all large. But it does help us to deduce other, far-reaching, properties such as the fact that the Legendre symbol is **strongly multiplicative.*** That is, for any integers m, n we have

$$\left(\frac{mn}{p}\right) = \left(\frac{m}{p}\right)\left(\frac{n}{p}\right), \tag{3.13}$$

(not only when $(m, n) = 1$). This is because each side of (3.13) must be either -1, 0 or 1; so if the two sides of (3.13) were different they would be incongruent modulo p. However, Euler's criterion implies that

$$\left(\frac{mn}{p}\right) \equiv (mn)^{\frac{1}{2}(p-1)} = m^{\frac{1}{2}(p-1)} \cdot n^{\frac{1}{2}(p-1)} \equiv \left(\frac{m}{p}\right) \cdot \left(\frac{n}{p}\right) \pmod{p},$$

whence (3.13) is true.

* Cf. equation (1.25).

Example Evaluate $(\frac{47}{17})$. $47 \equiv -4 \pmod{17}$ so by the properties already found we have

$$\left(\frac{47}{17}\right) = \left(\frac{-4}{17}\right) = \left(\frac{-1}{17}\right)\left(\frac{4}{17}\right) = \left(\frac{-1}{17}\right) = +1.$$

There are still some Legendre symbols, such as $(\frac{5}{17})$, which are difficult to evaluate by our present methods. The final tool, which will dispatch all obstinate cases, will be Gauss's famous law of quadratic reciprocity. Before proving it we shall need a lemma (or preliminary result) also due to Gauss. This will enable us to characterize, very effectively, the types of primes for which a given integer is a quadratic residue.

GAUSS'S LEMMA *Suppose that $a \not\equiv 0 \pmod{p}$. We know that the multiplies $1a, 2a, \ldots, \frac{1}{2}(p-1)a$ are all incongruent modulo p, so that their least positive remainders are $\frac{1}{2}(p-1)$ of the numbers $1, 2, \ldots, p-1$. If μ is the number of their remainders between $\frac{1}{2}(p+1)$ and $p-1$ inclusive, then*

$$\left(\frac{a}{p}\right) = (-1)^{\mu}. \qquad (3.14)$$

Proof Suppose the μ distinct remainders lying between $\frac{1}{2}(p+1)$ and $p-1$ are called $\alpha_1, \ldots, \alpha_{\mu}$; and that the other remainders, which lie between 1 and $\frac{1}{2}(p-1)$, are called $\beta_1, \ldots, \beta_{\nu}$. Then, since there are $\frac{1}{2}(p-1)$ remainders altogether, we must have $\mu + \nu = \frac{1}{2}(p-1)$. Consider now the set S consisting of the numbers $p - \alpha_1, \ldots, p - \alpha_{\mu}$, $\beta_1, \ldots, \beta_{\nu}$. S contains $\mu + \nu = \frac{1}{2}(p-1)$ numbers, each of which lies between 1 and $\frac{1}{2}(p-1)$ (certainly each β_j does; and for each α_i, $\frac{1}{2}(p+1) \leqslant \alpha_i \leqslant p-1$ implies $1 \leqslant p - \alpha_i \leqslant \frac{1}{2}(p-1)$). We shall show that the members of S are all distinct, which will mean that S must be the set of all consecutive numbers between 1 and $\frac{1}{2}(p-1)$ inclusive. All the $p - \alpha$'s are distinct since the α's are; also the β's are distinct. The only other possibility is that we might have $p - \alpha_i = \beta_j$ for some

α_i and β_j which are the remainders of, say, ar and as (with $1 \leqslant r \leqslant \frac{1}{2}(p - 1)$ and $1 \leqslant s \leqslant \frac{1}{2}(p - 1)$). Then $p = \alpha_i + \beta_j$, giving

$$0 \equiv \alpha_i + \beta_j \equiv a(r + s) \pmod{p}.$$

But this is impossible since $a \not\equiv 0 \pmod{p}$ and $r + s \not\equiv 0 \pmod{p}$ (as $2 \leqslant r + s \leqslant p - 1$). Thus no two of the numbers in S are the same, and as we said previously, this means that S consists of the numbers $1, \ldots, \frac{1}{2}(p - 1)$ in some order. Therefore, multiplying together all the numbers in S, we have*

$$\prod_{\substack{1 \leqslant i \leqslant \mu \\ 1 \leqslant j \leqslant \nu}} (p - \alpha_i)\beta_j = \tfrac{1}{2}(p - 1)!$$

Whence, taking congruences modulo p,

$$\prod_{\substack{1 \leqslant i \leqslant \mu \\ 1 \leqslant j \leqslant \nu}} (-\alpha_i)\beta_j = (-1)^\mu \prod_{\substack{1 \leqslant i \leqslant \mu \\ 1 \leqslant j \leqslant \nu}} \alpha_i\beta_j \equiv \tfrac{1}{2}(p - 1)! \pmod{p}.$$

Now, replacing the α's and β's by the original multiples a, $2a, \ldots, \frac{1}{2}(p - 1)a$, (to which they are congruent in some order) gives

$$(-1)^\mu a^{\frac{1}{2}(p-1)}(\tfrac{1}{2}(p - 1))! \equiv \tfrac{1}{2}(p - 1)! \pmod{p}$$

and so

$$(-1)^\mu a^{\frac{1}{2}(p-1)} \equiv 1 \pmod{p}.$$

From Euler's criterion this implies $(-1)^\mu \cdot \left(\dfrac{a}{p}\right) \equiv 1 \pmod{p}$, and since the left hand side of the last congruence is either $+1$ or -1 we can actually say $(-1)^\mu \cdot \left(\dfrac{a}{p}\right) = 1$. Multiplying both sides of this equality by $(-1)^\mu$ gives (3.14).

Example Use Gauss's lemma to find a simple rule for evaluating $\left(\dfrac{2}{p}\right)$. On replacing a by 2 in the statement of the lemma, we see that we must

---

* The notation $(\frac{1}{2}(p - 1))!$ means the product of all the natural numbers between 1 and $\frac{1}{2}(p - 1)$ inclusive.

determine how many of the multiples $1 . 2, \ldots, \frac{1}{2}(p - 1) . 2$ have remainders between $\frac{1}{2}(p + 1)$ and $p - 1$ inclusive. Since the multiples all lie between 1 and $p - 1$, they are their own least positive remainders, so we have to find how many of these even numbers $2x$ satisfy

$$\tfrac{1}{2}(p + 1) \leqslant 2x \leqslant p - 1. \tag{3.15}$$

Now for each odd prime p we have $(8, p) = 1$, so that $p \equiv -3, -1, 1$ or $3 \pmod 8$ since $\pm 3, \pm 1$ form a reduced set of residues modulo 8. In other words we can take $p = 8k \pm 3$ or $p = 8k \pm 1$ for some integer $k \geqslant 0$. Distinguishing between these cases, (3.15) becomes

$$
\begin{aligned}
4k - 1 &\leqslant 2x \leqslant 8k - 4 \text{ if } p = 8k - 3, \\
4k &\leqslant 2x \leqslant 8k - 2 \text{ if } p = 8k - 1, \\
4k + 1 &\leqslant 2x \leqslant 8k \qquad \text{if } p = 8k + 1, \\
4k + 2 &\leqslant 2x \leqslant 8k + 2 \text{ if } p = 8k + 3.
\end{aligned}
$$

The number μ of such x is odd in the first and last cases (being $2k - 1$ and $2k + 1$ respectively); and even—equal to $2k$—in both the second and third cases. Hence

$$\left(\frac{2}{p}\right) = (-1)^\mu := \begin{cases} +1 \text{ if } p \equiv \pm 1 \pmod 8 \\ -1 \text{ if } p \equiv \pm 3 \pmod 8. \end{cases} \tag{3.16}$$

Suppose now that we wish to evaluate $\left(\dfrac{n}{p}\right)$ for an odd prime p and a given integer $n \neq 0$; and that

$$n = (-1)^{e_0} . 2^{e_1} . q_2^{e_2} \ldots q_r^{e_r}. \tag{3.17}$$

In (3.17) the q's are distinct odd primes and the e's are non-negative integers ($e_i \geqslant 0$ for each $i, 0 \leqslant i \leqslant r$) with $e_0 = 0$ or 1 according as n is positive or negative. We shall assume that $p \nmid n$, or in other words that $p \neq q_i, 2 \leqslant i \leqslant r$; for otherwise we would know at once that $\left(\dfrac{n}{p}\right) = 0$. From the multiplicative property (3.13) we then have

$$\left(\frac{n}{p}\right) = \left(\frac{(-1)^{e_0}}{p}\right) \cdot \left(\frac{2^{e_1}}{p}\right) \cdot \left(\frac{q_2^{e_2}}{p}\right) \cdots \left(\frac{q_r^{e_r}}{p}\right)$$

$$= \left(\frac{-1}{p}\right)^{e_0} \cdot \left(\frac{2}{p}\right)^{e_1} \cdot \left(\frac{q_2}{p}\right)^{e_2} \cdots \left(\frac{q_r}{p}\right)^{e_r}. \qquad (3.18)$$

We can now immediately write down the values of $\left(\dfrac{-1}{p}\right)$ and $\left(\dfrac{2}{p}\right)$, and so immediately write down the values of $\left(\dfrac{-1}{p}\right)^{e_0}$ and $\left(\dfrac{2}{p}\right)^{e_1}$. Further, if for each i with $2 \leqslant i \leqslant r$ we knew the value of $\left(\dfrac{q_i}{p}\right)$, it would be easy to decide the value of $\left(\dfrac{q_i}{p}\right)^{e_i}$, and $\left(\dfrac{n}{p}\right)$ would then be determined by (3.18). So the only remaining uncertainty concerns the calculation of $\left(\dfrac{q}{p}\right)$ for any prime q different from p.

We shall investigate $\left(\dfrac{q}{p}\right)$ by using Gauss's lemma again, but in this case it will be helpful to have the notion of the 'integral part' of a number x which need not itself be an integer. The integral part of x, usually denoted by $[x]$, is defined as the largest integer not exceeding x. Alternatively we may say that $[x]$ is the unique integer satisfying

$$x - 1 < [x] \leqslant x. \qquad (3.19)$$

Examples   (i) $[\frac{1}{2}] = 0$;   (ii) $[5] = 5$;
            (iii) $[\pi] = 3$;   (iv) $[-7 \cdot 2] = -8$.

In order to examine $\left(\dfrac{q}{p}\right)$ using Gauss's lemma, we know that we must look at the least positive remainders, modulo p, of the multiples mq for $1 \leqslant m \leqslant \frac{1}{2}(p-1)$. As before, we shall let $\alpha_1, \ldots, \alpha_\mu$ be the names of the remainders between $\frac{1}{2}(p+1)$ and $p-1$; and $\beta_1, \ldots, \beta_\nu$ be the names of the remainders between 1 and $\frac{1}{2}(p-1)$. For each m with $1 \leqslant m \leqslant \frac{1}{2}(p-1)$ we have $mq = p \cdot \dfrac{mq}{p}$. Here $\dfrac{mq}{p}$ is not an integer (since $p \nmid m$, and $p \nmid q$), but the greatest integer below it is $\left[\dfrac{mq}{p}\right]$. Thus the largest integer multiple of p below mq is $p\left[\dfrac{mq}{p}\right]$, so we may write

$$mq = p\left[\frac{mq}{p}\right] + r_m \qquad (1 \leqslant r_m \leqslant p-1), \qquad (3.20)$$

where the remainder r_m is one of the α's if it is between $\frac{1}{2}(p+1)$ and $p-1$ inclusive, and one of the β's otherwise. Since each r_m is an α_i or a β_j we must have

$$\sum_{m=1}^{\frac{1}{2}(p-1)} r_m = \sum_{i=1}^{\mu} \alpha_i + \sum_{j=1}^{\nu} \beta_j.$$

Therefore, adding together the equations (3.20) for every m between 1 and $\frac{1}{2}(p-1)$, we obtain

$$q \sum_{m=1}^{\frac{1}{2}(p-1)} m = p \sum_{m=1}^{\frac{1}{2}(p-1)} \left[\frac{mq}{p}\right] + \sum_{i=1}^{\mu} \alpha_i + \sum_{j=1}^{\nu} \beta_j.$$

If we denote $\sum_{m=1}^{\frac{1}{2}(p-1)} \left[\dfrac{mq}{p}\right]$ by s, and use the fact that $\sum_{m=1}^{\frac{1}{2}(p-1)} m = \frac{1}{2}[\frac{1}{2}(p-1)][\frac{1}{2}(p+1)] = \frac{1}{8}(p^2-1)$, we then have

$$q[\tfrac{1}{8}(p^2-1)] = ps + \sum_{i=1}^{\mu} \alpha_i + \sum_{j=1}^{\nu} \beta_j. \qquad (3.21)$$

Now we know, from the proof of Gauss's lemma, that the numbers $p - \alpha_1, \ldots, p - \alpha_\mu, \beta_1, \ldots, \beta_\nu$ comprise the set

$\{1, \ldots, \frac{1}{2}(p - 1)\}$. So their sum, which is $\sum\limits_{i=1}^{\mu} (p - \alpha_i) + \sum\limits_{j=1}^{\nu} \beta_j$, must be equal to the sum of the first $\frac{1}{2}(p - 1)$ consecutive numbers, which is $\frac{1}{8}(p^2 - 1)$. That is,

$$\frac{1}{8}(p^2 - 1) = p\mu - \sum_{i=1}^{\mu} \alpha_i + \sum_{j=1}^{\nu} \beta_j. \tag{3.22}$$

Subtracting (3.22) from (3.21) now gives

$$\frac{1}{8}(q - 1)(p^2 - 1) = p(s - \mu) + 2\sum_{i=1}^{\mu} \alpha_i. \tag{3.23}$$

If (3.23) seems rather strange, it may be helpful to consider what happens when $q = 2$, for then (3.23) gives all our previous results about $\left(\dfrac{2}{p}\right)$. This is because, when $q = 2$, the multiples $2m$ $[1 \leqslant m \leqslant \frac{1}{2}(p - 1)]$ are their own least positive remainders, so that (3.20) just becomes $2m = r_m$. In other words, $\left[\dfrac{2m}{p}\right] = 0$ for each m, which means that $s = \sum\limits_{m=1}^{\frac{1}{2}(p-1)} \left[\dfrac{2m}{p}\right] = 0$. Therefore if $q = 2$ (and so $q - 1 = 1$), (3.23) becomes

$$\frac{1}{8}(p^2 - 1) = -p\mu + 2\sum_{i=1}^{\mu} \alpha_i. \tag{3.24}$$

On taking congruences modulo 2 we have

$\frac{1}{8}(p^2 - 1) \equiv -p\mu \equiv \mu \pmod{2}$(since $p \equiv -p \equiv 1 \pmod 2$)

Therefore,

$$\left(\frac{2}{p}\right) = (-1)^{\mu} = (-1)^{\frac{1}{8}(p^2 - 1)}. \tag{3.25}$$

Equation (3.25) is equivalent to (3.16), since $\frac{1}{8}(p^2 - 1)$ is even if $p \equiv \pm 1 \pmod 8$, and is odd if $p \equiv \pm 3 \pmod 8$.

When q is odd $q - 1$ is even, and it follows that (3.23) implies

$$0 \equiv p(s - \mu) \pmod 2$$
$$\equiv s - \mu \pmod 2.$$

Therefore $\mu \equiv s \pmod 2$ and we must have

$$\left(\frac{q}{p}\right) = (-1)^\mu = (-1)^s. \tag{3.26}$$

In view of the sum which s denotes, (3.26) is not as easy to work with as the corresponding result (3.25) in the case $q = 2$. However, (3.26) will be the key to the proof of Gauss's law of quadratic reciprocity—which is so called because it relates $\left(\frac{q}{p}\right)$ to $\left(\frac{p}{q}\right)$. Notice that if we let t stand for $\sum\limits_{m=1}^{\frac{1}{2}(q-1)} \left[\frac{mp}{q}\right]$, and reverse the roles of p and q in the argument leading to (3.23) and (3.26), we obtain

$$\left(\frac{p}{q}\right) = (-1)^t. \tag{3.27}$$

For convenience in proving the reciprocity property of the Legendre symbol we shall also set $p' = \frac{1}{2}(p - 1)$ and $q' = \frac{1}{2}(q - 1)$.

LAW OF QUADRATIC RECIPROCITY *If p and q are different odd primes, then*

$$\left(\frac{p}{q}\right) \cdot \left(\frac{q}{p}\right) = (-1)^{\frac{1}{2}(p-1) \cdot \frac{1}{2}(q-1)}; \tag{3.28}$$

or, multiplying both sides of (3.28) by $\left(\frac{q}{p}\right)$

$$\left(\frac{p}{q}\right) = (-1)^{p' \cdot q'} \left(\frac{q}{p}\right) \left(\text{because certainly } \left(\frac{q}{p}\right)^2 = 1\right).$$

At least one of p', q' is even if at least one of p or q is congruent to 1 modulo 4. So the result can also be stated as

$$\left(\frac{p}{q}\right) = \begin{cases} \left(\dfrac{q}{p}\right) & \text{if either } p \equiv 1 \text{ (mod 4) or } q \equiv 1 \text{ (mod 4)} \\ -\left(\dfrac{q}{p}\right) & \text{if both } p \equiv -1 \text{ (mod 4) and } q \equiv -1 \text{ (mod 4).} \end{cases}$$

(3.29)

Proof From (3.26) and (3.27) we have $\left(\dfrac{p}{q}\right) \cdot \left(\dfrac{q}{p}\right) = (-1)^{s+t}$,
so it will suffice to prove that $s + t = p'q'$. We shall do this by
counting, in two different ways, the number of members of the
set V of integers each of which can be expressed as $qx - py$,
for some integers x and y with $1 \leqslant x \leqslant p'$ and $1 \leqslant y \leqslant q'$.

First of all, since the x's come from a reduced set of residues
modulo p, and the y's from a reduced set of residues modulo q;
the numbers $qx - py$ must be $p'q'$ members of a reduced set of
residues* modulo pq. That is, they must all be incongruent
modulo pq, and none of them is divisible by pq. So, in
particular, V contains $p'q'$ distinct numbers, none of which
are zero.

On the other hand, for each x with $1 \leqslant x \leqslant p'$, $\left[\dfrac{qx}{p}\right]$
is the number of natural numbers less than $\dfrac{qx}{p} < \dfrac{q}{2}$. In other
words, $\left[\dfrac{qx}{p}\right]$ is the number of ways of choosing an integer y,
say, $[1 \leqslant y \leqslant \frac{1}{2}(q - 1)]$ such that $y < \dfrac{qx}{p}$, or equivalently,
$qx - py > 0$. But this is the number of positive members of V
which can be formed using the given x. So the total number of
positive numbers in V is $\sum_{x=1}^{p'} \left[\dfrac{qx}{p}\right] = s$. Similarly the number

---

* See the proof (in Chapter 2) of the fact that ϕ is a multiplicative function.

of negative numbers in V is $\sum_{y=1}^{q'} \left[\dfrac{py}{q} \right] = t$. So $p'q'$, the total number of members of V, must be equal to $s + t$, and this completes the proof.

Example 1 Evaluate $(\frac{5}{17})$. Using quadratic reciprocity we have

$$(\tfrac{5}{17}) = (\tfrac{17}{5}) = (\tfrac{2}{5}) = -1.$$

Example 2 Determine whether or not the congruence $x^2 \equiv 521 \pmod{997}$ is soluble. Here 521 and 997 are both primes which are congruent to 1 modulo 4, so we obtain

$(\tfrac{521}{997}) = (\tfrac{476}{521}) = (\tfrac{4}{521})(\tfrac{7}{521})(\tfrac{17}{521}) = (\tfrac{7}{521})(\tfrac{17}{521})$ (since 4 is a square)

$\quad = (\tfrac{521}{7})(\tfrac{521}{17}) = (\tfrac{3}{7})(\tfrac{11}{17}) = -(\tfrac{7}{3})(\tfrac{17}{11})$

$\quad = -(\tfrac{1}{3})(\tfrac{6}{11}) = -(\tfrac{2}{11})(\tfrac{3}{11}) = (\tfrac{3}{11})$

$\quad = -(\tfrac{11}{3}) = -(\tfrac{2}{3}) = +1.$

Hence the congruence is soluble.

Example 3 Is the congruence $x^2 + 6x - 5 \equiv 0 \pmod{127}$ soluble? Completing the square gives $(x + 3)^2 \equiv 14 \pmod{127}$ and

$$(\tfrac{14}{127}) = (\tfrac{2}{127})(\tfrac{7}{127}) = (\tfrac{7}{127}) = -(\tfrac{127}{7}) = -(\tfrac{1}{7}) = -1.$$

So the congruence is not soluble.

Of course, once we know that a particular congruence is soluble, we still have the problem of finding the solutions. Quite often in applications though, it is sufficient to know that a given congruence is soluble without actually knowing what the solutions are. This will be the case in Chapter 4 when we use our knowledge of quadratic congruences to study sums of squares.

EXERCISES

1. If p is a prime and d is a natural number which divides $p - 1$, then we can write $x^{p-1} - 1 = (x^d - 1)(x^{p-1-d} + \ldots + 1)$. Use this identity to deduce that, for every divisor d of $p - 1$, the congruence $x^d \equiv 1 \pmod{p}$ has exactly d solutions in any complete set of residues modulo p.

2. Let p be a prime. Each number $a \in \{1, \ldots, p-1\}$ can be paired with another number $a' \in \{1, \ldots, p-1\}$ so that $aa' \equiv 1 \pmod{p}$. Find conditions which ensure that such numbers a, a' are distinct. By similarly pairing off as many as possible of the numbers $1, \ldots, p-1$, prove Wilson's* theorem that $(p-1)! \equiv -1 \pmod{p}$.

3. Prove that if n is a composite number then $(n-1)! \equiv 0 \pmod{n}$, apart from $n = 4$ when $3! \equiv 2 \pmod 4$.

4. Show that 3 is a primitive root modulo 17, and use it to construct a table of indices modulo 17.

5. Using the table constructed in question 4, solve the following congruences:

   (i) $3x \equiv 7 \pmod{17}$; (ii) $3x^2 \equiv 11 \pmod{17}$; (iii) $8x^5 \equiv 14 \pmod{17}$.

6. Use Gauss's lemma to determine the primes for which 3 is a quadratic residue.

7. In each of the following cases, how many of the numbers in a reduced set of residues modulo p are kth power residues modulo p:

   (i) $k = 3, p = 19$; (ii) $k = 5, p = 23$; (iii) $k = 6, p = 29$.

8. Are the following congruences soluble?

   (i) $x^2 + 5x + 1 \equiv 0 \pmod{41}$; (ii) $5x^2 + 7x + 1 \equiv 0 \pmod{31}$;
   (iii) $x^2 + 6x - 154 \equiv 0 \pmod{399}$.

* 1741–93.

CHAPTER FOUR

Sums of Integer Sequences

1. Introduction

In this chapter we shall use our accumulated tools to investigate those numbers which can be characterized as sums of particular types of integers, such as squares or primes. Special cases of this kind of problem were originally discussed by Fermat, Legendre, Lagrange, Waring (1734–98), and Goldbach (1690–1764). In the present century, their investigations have broadened into a large field of study usually called **additive number theory.** We shall first examine sums of squares since the complete answers are fairly accessible in this case.

2. Sums of two squares

A classical problem, which we mentioned at the beginning of the book, is to find conditions which ensure that a given natural number n can be represented as the sum of at most two square integers. Since 0 is the square of an integer, we allow the possibility that n is representable by just one positive square (i.e. $n = x^2 + 0^2$ for some integer x). An essential aid in simplifying this problem is the fact that if two numbers are each the sum of two squares, then their product is also. This is because for any integers x_1, x_2, y_1, y_2 we have:

$$(x_1^2 + x_2^2)(y_1^2 + y_2^2) = (x_1 y_1 + x_2 y_2)^2 + (x_1 y_2 - x_2 y_1)^2.$$

(4.1)

69

The identity (4.1) suggests that we should first examine the multiplicative building blocks of the natural numbers: namely, the prime numbers. For if we knew which prime numbers were representable as a sum of two squares, then we could multiply the representable primes together and obtain (by (4.1)) other natural numbers which are sums of two squares. Now the prime 2 is easy since $2 = 1^2 + 1^2$; and for an odd prime p, Fermat found the simple criterion for p to be a sum of two squares. In giving Fermat's result we shall use his own method of proof (which he called his method of descent) since we shall need it again in discussing sums of four squares.

THEOREM   *Every prime p which is congruent to 1 modulo 4 can be expressed as a sum of two squares.*

Proof   From Chapter 3 we know that -1 is a quadratic residue of such a p, so there must be an integer x_0 such that

$$x_0^2 \equiv -1 \,(\mathrm{mod}\, p)$$

or

$$x_0^2 + 1^2 \equiv 0 \,(\mathrm{mod}\, p).$$

We can choose x_0 to be in the complete set of residues

$$-\tfrac{1}{2}(p - 1), \ldots, -1, 0, \ldots \tfrac{1}{2}(p - 1);$$

so that $|x_0| < \dfrac{p}{2}$, and consequently

$$1 \leqslant x_0^2 + 1 < 1 + \left(\frac{p}{2}\right)^2 < p^2.$$

There is therefore some multiple of p which is less than p^2 and which is expressible as a sum of two positive squares. In particular, the smallest positive multiple of p which is so expressible is less than p^2. Our object is to prove that the least positive multiple of p which can be expressed as a sum of two squares must be p itself. This will follow when we show that no multiple greater than p (and less than p^2) can be the

70

least. We shall do this by showing that if mp ($1 < m < p$) is a sum of two squares then we can always point to a positive multiple $lp < mp$ which is also a sum of two squares. Suppose then that

$$mp = x_1^2 + x_2^2, \qquad (4.2)$$

where $1 < m < p$, and x_1, x_2 are integers which are not both zero. Now we can choose integers y_1, y_2 such that

$$y_1 \equiv x_1 \pmod{m} \quad \text{and} \quad |y_1| \leqslant \frac{m}{2},$$
$$\qquad (4.3)$$
$$y_2 \equiv x_2 \pmod{m} \quad \text{and} \quad |y_2| \leqslant \frac{m}{2},$$

because we can take y_1, y_2 to be in a complete set of residues between $-\dfrac{m}{2}$ and $\dfrac{m}{2}$. Therefore, using (4.2),

$$y_1^2 + y_2^2 \equiv x_1^2 + x_2^2 \equiv 0 \pmod{m},$$

so that $\qquad y_1^2 + y_2^2 = lm$ for some integer l. $\qquad (4.4)$

In (4.4), $|y_1| \leqslant \dfrac{m}{2}$ and $|y_2| \leqslant \dfrac{m}{2}$ imply $lm \leqslant \dfrac{m^2}{4} + \dfrac{m^2}{4} = \dfrac{m^2}{2}$

and so $l \leqslant \dfrac{m}{2} < m$. Also, since $y_1^2 + y_2^2$ is never negative, we have at once $l \geqslant 0$. We can actually rule out the possibility that $l = 0$, because, from (4.4), $l = 0$ would imply $y_1 = y_2 = 0$. From (4.3), this would mean $x_1 \equiv x_2 \equiv 0 \pmod{m}$. Then $x_1^2 + x_2^2$ would be a multiple of m^2 whence, from (4.2), p would be a multiple of m. But this is not possible as $1 < m < p$. Therefore we must have

$$1 \leqslant l < m. \qquad (4.5)$$

Now multiplying together the equations (4.2) and (4.4), and

71

using (4.1), we have

$$lpm^2 = (x_1^2 + x_2^2)(y_1^2 + y_2^2)$$
$$= (x_1y_1 + x_2y_2)^2 + (x_1y_2 - x_2y_1)^2.$$

From (4.2) and (4.3), $x_1y_1 + x_2y_2 \equiv x_1^2 + x_2^2 \equiv 0 \pmod{m}$ and $x_1y_2 - x_2y_1 \equiv x_1x_2 - x_2x_1 \equiv 0 \pmod{m}$. So we can divide both sides of the last equality by m^2 and obtain a representation of lp as a sum of two integral squares:

$$lp = \left(\frac{x_1y_1 + x_2y_2}{m}\right)^2 + \left(\frac{x_1y_2 - x_2y_1}{m}\right)^2.$$

Inequality (4.5) implies that lp is a positive multiple of p which is less than mp, so the proof is complete.

It is easy to show that no prime $q \equiv -1 \equiv 3 \pmod{4}$ can be a sum of two integer squares, because every square is congruent to either 0 or 1 modulo 4, so that a sum of two squares can only be congruent to 0, 1, or 2 modulo 4. However, any even power of such a q is certainly a sum of two squares, since for every $s \geqslant 0$ we have $q^{2s} = (q^s)^2 + 0^2$. We can now deduce that

the only natural numbers which are sums of two integer squares are products of: powers of 2, powers of primes $p \equiv 1 \pmod{4}$, and even powers of primes $q \equiv -1 \pmod{4}$. (4.6)

Using (4.1) we can see that any number of this form is a sum of two squares. On the other hand, the first number which is a sum of two squares—namely $1 = 1^2 + 0^2$—is of the above form (with all exponents equal to zero). So we may assume inductively that every sum of two squares which is less than $n > 1$ is of the stated form, and that

$$n = x^2 + y^2 \tag{4.7}$$

for some integers x, y. We must prove that n is of the form given in (4.6). This is certainly true if n is not divisible by any prime congruent to -1 modulo 4. So suppose that $q \mid n$ for some prime number $q \equiv -1 \pmod 4$. Then $x^2 + y^2 \equiv 0 \pmod q$, or

$$x^2 \equiv -y^2 \pmod q. \tag{4.8}$$

Now if $y \not\equiv 0 \pmod q$ (that is, $(y, q) = 1$), there would be an integer z such that $yz \equiv 1 \pmod q$. Whence multiplying both sides of (4.8) by z^2 would give $(xz)^2 \equiv -1 \pmod q$. This would mean $\left(\dfrac{-1}{q}\right) = +1$ which we know is false; so in (4.8) we must have $y \equiv 0 \pmod q$ and thus $x \equiv 0 \pmod q$ also. If, say, $x = qx_1$ and $y = qy_1$, then (4.7) becomes

$$n = q^2(x_1^2 + y_1^2),$$

where $x_1^2 + y_1^2 = \dfrac{n}{q^2} < n$. Therefore, by the inductive hypothesis, $x_1^2 + y_1^2$ is of the form stated in (4.6). Hence n is also of the required form, so our proof by induction is complete.

Example 1 $5 = 1^2 + 2^2$ and $13 = 2^2 + 3^2$ so (4.1) implies
$$\begin{aligned}
65 &= (1^2 + 2^2)(2^2 + 3^3) \\
&= (1 . 2 + 2 . 3)^2 + (1 . 3 - 2 . 2)^2 \\
&= 8^2 + (-1)^2 = 8^2 + 1^2.
\end{aligned}$$

Example 2   The following numbers are each of the form stated in (4.6) and can therefore be represented as a sum of two squares:

        (i) 997, which is a prime $\equiv 1 \pmod 4$;
        (ii) $90 = 2 . 3^2 . 5$;
        (iii) $221 = 13 . 17$.

Example 3   The following numbers are not of the form specified by (4.6) and so cannot be represented as a sum of two squares:

        (i) 47;   (ii) $126 = 2 . 3^2 . 7$;
        (iii) $60 = 2^2 . 3 . 5$;   (iv) $21 = 3 . 7$.

3. Sums of three squares

Having completely settled the problem of which numbers are sums of two squares, it is natural to try to extend this study by considering sums of three squares. So we ask: Can we find simple conditions which, when satisfied, ensure that a given number is a sum of at most three squares? If we adopt an experimental approach to this question, we find that most numbers can be represented by a sum of three squares, the first few exceptions being $7, 15, 23, \ldots$. These first exceptions are all congruent to 7 modulo 8, and it is easy to show that no number congruent to 7 modulo 8 can be a sum of three squares. This is because every integer square is congruent to 0, 1 or 4 (mod 8), so (just by trying each possibility) we see that a sum of three squares can never be 7 (mod 8). Indeed we can go further and prove that if $h \geqslant 0$ and $k \geqslant 0$ are integers, then **no number of the form $4^h(8k + 7)$ can be a sum of three squares.** We know that this is true if $h = 0$, so suppose that $h \geqslant 1$ and that we had

$$4^h(8k + 7) = x_1^2 + x_2^2 + x_3^2 \qquad (4.9)$$

for some integers x_1, x_2, x_3. Since $x_i^2 \equiv 0$ (mod 4) if $2 \mid x_i$, and $x_i^2 \equiv 1$ (mod 4) if $2 \nmid x_i$, we see that the only way $x_1^2 + x_2^2 + x_3^2$ could be a multiple of 4 would be for each x_i to be even. Whence we would have

$$4^{h-1}(8k + 7) = \left(\frac{x_1}{2}\right)^2 + \left(\frac{x_2}{2}\right)^2 + \left(\frac{x_3}{2}\right)^2.$$

By repeatedly dividing by 4, we would obtain after h steps

$$8k + 7 = \left(\frac{x_1}{2^h}\right)^2 + \left(\frac{x_2}{2^h}\right)^2 + \left(\frac{x_3}{2^h}\right)^2,$$

with $\frac{x_1}{2^h}, \frac{x_2}{2^h}, \frac{x_3}{2^h}$ all being integers. Since $8k + 7$ cannot be a sum of three squares we can conclude as desired that there are no integers x_1, x_2, x_3 which satisfy (4.9).

The result we have just proved forms one half of the complete answer to the question of which numbers are sums of three squares. For it is also known that **if a number is not of the form $4^h(8k + 7)$ then it can be represented by a sum of three integer squares.** This last result was first proved by Legendre in 1798, but the proof involves the theory of quadratic forms so we shall not give it here. However, it should not come as a surprise that the proof of the three-square result is of a different character from that for two squares. For in the case of sums of three squares there is no formula which corresponds to (4.1). Thus it is not always true that if each of two numbers is a sum of three squares then their product is also. For example, $3 = 1^2 + 1^2 + 1^2$ and $5 = 2^2 + 1^2 + 0^2$, but $3 \cdot 5 = 15$ cannot be expressed as a sum of three squares.

4. Sums of four squares

The question of which numbers can be represented by a sum of more than three squares is completely answered by

LAGRANGE'S THEOREM   *Every natural number can be expressed as a sum of four integer squares.*

Proof   Our proof of Lagrange's theorem will be guided by our treatment of two squares, because for four squares we have the following analogue of (4.1):

$$(x_1^2 + x_2^2 + x_3^2 + x_4^2)(y_1^2 + y_2^2 + y_3^2 + y_4^2)$$
$$= (x_1 y_1 + x_2 y_2 + x_3 y_3 + x_4 y_4)^2$$
$$+ (x_1 y_2 - x_2 y_1 + x_3 y_4 - x_4 y_3)^2 \quad (4.10)$$
$$+ (x_1 y_3 - x_3 y_1 - x_2 y_4 + x_4 y_2)^2$$
$$+ (x_1 y_4 - x_4 y_1 - x_3 y_2 + x_2 y_3)^2.$$

So if we show that each prime can be written as a sum of at

most four squares, then (4.10) will immediately imply that this is also true of every natural number. We already know that $2 = 1^2 + 1^2$, and that each prime which is congruent to 1 modulo 4 is a sum of two squares. Therefore all that remains to be shown is that each prime $q \equiv -1 \pmod 4$ can be represented as a sum of four squares.

As in the proof of Fermat's result about primes which are congruent to 1 modulo 4, we shall first prove that some multiple of q is a sum of four squares. In order to see this we argue as follows. The number 1 is certainly a quadratic residue modulo q. So if we consider in turn the numbers $2, 3, \ldots, q - 1$, we shall come across a first quadratic non-residue, say $a + 1$ for some $a > 0$. That is,

$$\left(\frac{a + 1}{q}\right) = -1,$$

and the preceding number a will then be a quadratic residue:

$$\left(\frac{a}{q}\right) = +1.$$

Therefore, since $q \equiv -1 \pmod 4$ and so $\left(\frac{-1}{q}\right) = -1$, we have

$$\left(\frac{-a - 1}{q}\right) = \left(\frac{-1}{q}\right)\left(\frac{a + 1}{q}\right) = +1.$$

In other words there must be integers x_1, x_2 such that $x_1^2 \equiv a \pmod q$ and $x_2^2 \equiv -a - 1 \pmod q$. Adding these congruences together then gives

$$x_1^2 + x_2^2 + 1^2 + 0^2 \equiv 0 \pmod q.$$

Also x_1, x_2 can both be chosen to lie between $-\frac{q}{2}$ and $\frac{q}{2}$ so that $q \leqslant 1 + x_1^2 + x_2^2 < 1 + 2\left(\frac{q}{2}\right)^2 < q^2$. So there is some multiple of q, less than q^2, which is a sum of four squares.

The fact that q itself is the least such multiple will again follow from Fermat's idea of descent from a given representation as a sum of squares to a smaller one. Thus suppose that for some m with $1 < m < q$ we have

$$mq = x_1^2 + x_2^2 + x_3^2 + x_4^2, \qquad (4.11)$$

with x_1, x_2, x_3, x_4 being integers. We shall show that (4.11) necessarily leads to a representation of a smaller multiple of q as a sum of four squares. First suppose that m is even. Then, in order for both sides of (4.11) to be even, we must have either no odd x's, two odd x's, or four odd x's. Whence, by relabelling the x's if necessary, we may assume that $x_1 \equiv x_2$ (mod 2) and $x_3 \equiv x_4$ (mod 2). This means that $\frac{1}{2}(x_1 \pm x_2)$ and $\frac{1}{2}(x_3 \pm x_4)$ are integers, so that we have the following representation of the smaller multiple $\dfrac{m}{2} \cdot q$:

$$\frac{m}{2} \cdot q = [\tfrac{1}{2}(x_1 + x_2)]^2 + [\tfrac{1}{2}(x_1 - x_2)]^2 + [\tfrac{1}{2}(x_3 + x_4)]^2$$

$$+ [\tfrac{1}{2}(x_3 - x_4)]^2. \qquad (4.12)$$

When m is odd we can use exactly the same method as in our discussion of sums of two squares. That is, we choose integers y_1, y_2, y_3, y_4 such that for each i with $1 \leqslant i \leqslant 4$ we have $y_i \equiv x_i$ (mod m) and $-\dfrac{m}{2} < y_i < \dfrac{m}{2}$ (we can be sure there is strict inequality since $\dfrac{m}{2}$ is not an integer). Then using (4.11)

$$y_1^2 + y_2^2 + y_3^2 + y_4^2 \equiv x_1^2 + x_2^2 + x_3^2 + x_4^2 \equiv 0 \ (\text{mod } m),$$

so that

$$y_1^2 + y_2^2 + y_3^2 + y_4^2 = lm \text{ for some integer } l \geqslant 0. \quad (4.13)$$

Just as in (4.4) we must actually have $l \geqslant 1$, because $l = 0$ would imply that, for each i, $y_i = 0$ and thus $x_i \equiv 0$ (mod m).

Whence dividing both sides of (4.11) by m would mean that the prime q would be a multiple of m, and this is impossible as $1 < m < q$. Also, since $|y_i| < \dfrac{m}{2}$ for each i, we have $y_i^2 < \dfrac{m^2}{4}$, so from (4.13) $lm < m^2$ and $l < m$. Now, multiplying together the representations of mq and lm in (4.11) and (4.13), we see that lm^2q can be expressed as the sum of the four integral squares on the right hand side of (4.10). But since, for each i, $x_i \equiv y_i \pmod{m}$, we must have $x_1y_1 + x_2y_2 + x_3y_3 + x_4y_4 \equiv x_1^2 + x_2^2 + x_3^2 + x_4^2 \equiv 0 \pmod{m}$, and each of the expressions $x_1y_2 - x_2y_1 + x_3y_4 - x_4y_3, x_1y_3 - x_3y_1 - x_2y_4 + x_4y_2$ and $x_1y_4 - x_4y_1 - x_3y_2 + x_2y_3$ must also be congruent to zero modulo m. We can therefore divide by m^2 on both sides of the expression for lm^2q as a sum of four squares and obtain lq as the following sum of four integral squares:

$$lq = \left(\frac{x_1y_1 + x_2y_2 + x_3y_3 + x_4y_4}{m}\right)^2$$
$$+ \left(\frac{x_1y_2 - x_2y_1 + x_3y_4 - x_4y_3}{m}\right)^2$$
$$+ \left(\frac{x_1y_3 - x_3y_1 - x_2y_4 + x_4y_2}{m}\right)^2$$
$$+ \left(\frac{x_1y_4 - x_4y_1 - x_3y_2 + x_2y_3}{m}\right)^2.$$

This finishes the process of descent, and so completes the proof of Lagrange's theorem, because lq is a smaller positive multiple of q than mq.

5. Sums of other sequences

Questions similar to those discussed above for sums of squares can also be asked about other sequences such as cubes or primes; or indeed any given sequence A of non-negative integers a_1, a_2, a_3, ... For example, we can ask if every natural number can be represented as a sum of not more than a fixed number, say h, of elements of A. If this is so then we shall say that A is a **basis** (for **N**) of order h. If the smallest positive number in A is greater than 1 then every sum of elements of A will either be zero or greater than 1; so we can only hope for an affirmative answer to our question if we consider sequences which include 1. Also, in order to be able to say that 1, for example, is a sum of exactly h members of A, we shall always suppose that $0 \in A$.

In this wider context, Lagrange's theorem of the preceding section becomes the statement that the sequence of squares 0, 1, 4, ..., n^2, ... is a basis of order 4. It is now natural to enquire whether or not sequences of higher powers also form bases. That is, if $k > 2$ is a natural number, is there a number h (depending only on k) such that the sequence of kth powers 0, 1, 2^k, ..., n^k, ... is a basis of order h? Waring, in 1770, first speculated that this is always so. But it was not until 1909 that Hilbert, using very complicated analytic methods, first proved Waring's conjecture for all values of k.

The sequence of cubes furnishes a good example here of an interesting difference in behaviour between squares and higher powers. In this case, the known result which corresponds to Lagrange's theorem is the fact that the sequence of non-negative cubes 0, 1, 8, 27, ... is a basis of order 9. But actually only two numbers, 23 and 239, need 9 cubes to represent them: all others can be expressed as a sum of 8 or fewer cubes. Indeed the slightly weaker result that all numbers greater than $2 . 10^{23}$ are sums of 8 cubes,* can be proved using nothing more advanced than Legendre's three-square theorem. This is in complete contrast to the situation for sums of squares, since we know from Section 3 that there are infinitely many numbers which require the maximum number of squares to

* See G. L. Watson in *Mathematical Gazette*, vol. 37 (1953) pp. 209–11.

represent them. It is also known that all sufficiently large numbers can be represented by sums of 7 cubes, but in this case no estimate is known for the size of the largest number which requires 8 cubes. The behaviour of kth powers for $k > 3$ is similar. That is, there is a number $g(k)$ such that every natural number can be expressed as a sum of $g(k)$ kth powers. But only a finite number of positive integers require as many as $g(k)$ summands. All sufficiently large numbers can be represented as sums of some minimum number, $G(k) < g(k)$, of kth powers. Thus $g(2) = G(2) = 4$, whereas $g(3) = 9$ and $G(3) \leqslant 7$. The exact value of $G(k)$ is unknown for most values of k.

Sums of many other particular types of numbers have also been investigated; but often with special methods designed for the sequences in question, and often, too, with incomplete success. So although questions in this area can often be very difficult, there are undoubtedly many more beautiful results awaiting discovery.

EXERCISES

1. Express each of the following numbers as sums of two squares:

   (i) 997,　　　(ii) 90,　　　(iii) 221.

2. Use the facts that $13 = 2^2 + 3^2$ and $41 = 4^2 + 5^2$ to express $533 = 13 . 41$ as a sum of two squares in two different ways.

3. Use as few summands as possible to express each of the following numbers as a sum of squares:

   (i) 60,　　　(ii) 85,　　　(iii) 21,　　　(iv) 361,　　　(v) 47.

4. Prove that every integer can be expressed as $n_1^2 - n_2^2 - n_3^3$ for integers n_1, n_2, n_3. Find a number which needs three non-zero terms.

5. Use Legendre's three-square theorem to deduce that every natural number of the form $8n + 3$ is a sum of three odd squares.

6. Use the result of question 5 to deduce that every positive odd multiple of 4 can be expressed as a sum of four odd squares.

7. The n'th *triangular number* is the sum of the numbers from 1 to n inclusive, namely $\frac{1}{2}n(n + 1)$. Use the result of question 5 to deduce that every natural number is a sum of at most three triangular numbers.

8. Express 23 and 239 as sums of 9 cubes.

9. Prove that every integral cube is congruent to 0 or ± 1 (mod 9). By considering numbers of the form $9m \pm 4$, deduce that $G(3) \geqslant 4$.

10. Goldbach conjectured in 1742 that every even number greater than 2 is a sum of two primes. It is still not known whether this is always true; but prove that, if true, it would imply that the sequence $0, 1, 2, 3, 5, 7, 11, \ldots,$ $p, \ldots$ is a basis of order 3.

Appendix

Proof that primitive roots exist for a given modulus m if and only if

$$m = 1, 2, 4, p^\alpha \text{ or } 2p^\alpha, \text{ where } p \text{ is an odd prime and } \alpha \geqslant 1.$$

$$\text{(A.1)}$$

I. There are primitive roots modulo m in each of the cases (A.1)

(i) $m = 1, 2, 4$. Here primitive roots are 1, 1, 3 respectively.

(ii) $m = p$, an odd prime. Let d be any divisor of $p - 1$ and suppose that N_d is the total number of integers (in a reduced set of residues modulo p) which have order d. If there are no such integers then $N_d = 0$. Otherwise there is at least one such integer, a, say. Then each of the d incongruent powers a^h, with $0 \leqslant h < d$, is a solution of

$$x^d \equiv 1 \pmod{p} \qquad \text{(A.2)}$$

since $(a^h)^d = (a^d)^h \equiv 1 \pmod{p}$. Indeed these powers a^h are all the solutions of (A.2) because the congruence has exactly d incongruent solutions (see Exercise 1, Chapter 3). The N_d numbers of order d must all therefore be found among the powers a^h with $0 \leqslant h < d$. But if $(h, d) = k > 1$ then $(a^h)^{d/k} = (a^d)^{h/k} \equiv 1 \pmod{p}$ so such an a^h could not be of order d. Hence the only numbers which could be of order d are the

81

$\phi(d)$ numbers a^h with $(h, d) = 1$. Thus, whether $N_d = 0$ or $N_d > 0$, we certainly have $N_d \leqslant \phi(d)$.

On the other hand, the numbers between 1 and $p - 1$ inclusive form a reduced set of residues modulo p, so each of them has some order which is a divisor of $\phi(p) = p - 1$. That is to say $\sum_{d \mid p-1} N_d = p - 1$; and also, from (2.19) $\sum_{d \mid p-1} \phi(d) = p - 1$. Now $N_d \leqslant \phi(d)$ for every d, so if, for even one d, we had $N_d < \phi(d)$, then we would have the contradiction $\sum_{d \mid p-1} N_d < \sum_{d \mid p-1} \phi(d)$. This establishes that $N_d = \phi(d)$ for each d. In particular, the number of primitive roots is $N_{p-1} = \phi(p - 1) \geqslant 1$.

(iii) $m = p^\alpha$, $\alpha > 1$. Suppose first that g is any primitive root modulo p: say $g^{p-1} = 1 + wp$. Then for any natural number t, $g + tp$ is also a primitive root modulo p, and*

$$(g + tp)^{p-1} = g^{p-1} + (p - 1)tp \cdot g^{p-2} + (tp)^2 K$$

$$\text{for some integer } K$$

$$= 1 + wp - tp \cdot g^{p-2} + p^2 K_1$$

$$\text{for some integer } K_1$$

$$= 1 + p(w - t \cdot g^{p-2} + pK_1).$$

Now $g^{p-2} \not\equiv 0 \pmod p$, so, by taking $t = t_1$ for a suitable t_1, we can have $(g + t_1 p)^{p-1} = 1 + pu_1$ where $p \nmid u_1$. Thus, for $\alpha > 1$, we can assume inductively that $g + t_1 p$ is a primitive root modulo $p^{\alpha-1}$, and that $(g + t_1 p)^{\phi(p^{\alpha-1})} = 1 + p^{\alpha-1} u_{\alpha-1}$ where $p \nmid u_{\alpha-1}$. Then if d is the order of $g + t_1 p$ modulo p^α we have

$$(g + t_1 p)^d \equiv 1 \pmod{p^\alpha} \text{ so that certainly}$$

$$(g + t_1 p)^d \equiv 1 \pmod{p^{\alpha-1}}.$$

---

* The expansions used here are particular instances of expansions such as $(a + b)^n = a^n + na^{n-1}b + b^2 K$ ($n \geqslant 2$) which may be easily verified by induction on n without using the general binomial theorem.

This implies that d is divisible by $\phi(p^{\alpha-1}) = p^{\alpha-2}(p-1)$ (since $g + t_1 p$ is a primitive root modulo $p^{\alpha-1}$). Also, d must divide $\phi(p^\alpha) = p^{\alpha-1}(p-1)$, whence it must be either

$$p^{\alpha-2}(p-1) \quad \text{or} \quad p^{\alpha-1}(p-1).$$

But from the inductive hypotheses

$$(g + t_1 p)^{p^{\alpha-2}(p-1)} = 1 + p^{\alpha-1}u_{\alpha-1}$$
$$\not\equiv 1 \pmod{p^\alpha} \text{ (because } p \nmid u_{\alpha-1}).$$

Therefore $d = p^{\alpha-1}(p-1)$ and $g + t_1 p$ is a primitive root modulo p^α. Further,

$$(g + t_1 p)^{\phi(p^\alpha)} = (g + t_1 p)^{p^{\alpha-1}(p-1)} = (1 + p^{\alpha-1}u_{\alpha-1})^p$$
$$= 1 + p^\alpha u_{\alpha-1} + \tfrac{1}{2}p(p-1)p^{2\alpha-2}u_{\alpha-1}^2 + p^{3\alpha-3}K$$
$$\text{for some integer } K,$$

$$= 1 + p^\alpha u_{\alpha-1} + p^{\alpha+1}K_1 \quad \text{for some integer } K_1,$$
$$= 1 + p^\alpha(u_{\alpha-1} + pK_1)$$
$$= 1 + p^\alpha u_\alpha$$

where $u_\alpha = u_{\alpha-1} + pK_1$ is not divisible by p since $p \nmid u_{\alpha-1}$.

Therefore the truth of the inductive hypotheses for p^α follows from their truth for $p^{\alpha-1}$. Since they are true for $\alpha - 1 = 1$ they must hold for all $\alpha > 1$. In particular, the fixed number $g + t_1 p$ is a primitive root for all powers of p.

(iv) $m = 2p^\alpha$, $\alpha \geqslant 1$. Let now g be a primitive root modulo p^α and let g' be the odd one of g, $g + p^\alpha$. Then g' is a primitive root modulo p^α and $(g', 2p^\alpha) = 1$. If d is the order of g' modulo $2p^\alpha$ then $d \mid \phi(2p^\alpha) = \phi(2) \cdot \phi(p^\alpha) = \phi(p^\alpha)$. But $(g')^d \equiv 1 \pmod{2p^\alpha}$ implies $(g')^d \equiv 1 \pmod{p^\alpha}$ whence $\phi(p^\alpha) \mid d$, since g' is of order $\phi(p^\alpha)$ modulo p^α. Therefore $d = \phi(p^\alpha) = \phi(2p^\alpha)$ and g' is a primitive root modulo $2p^\alpha$.

II. There are no primitive roots modulo m if m is not given by (A.1)

This is because any m, not given by (A.1), comes under one of the following two cases.

(i) $m = 2^\alpha$, $\alpha \geq 3$. Here $\phi(m) = 2^{\alpha-1}$, and if $(a, 2) = 1$ we have $a^2 \equiv 1 \pmod 8$ so we may suppose inductively that for $\alpha \geq 3$, $a^{2^{\alpha-2}} \equiv 1 \pmod{2^\alpha}$. Then

$$a^{2^{\alpha-2}} = 1 + 2^\alpha t$$

whence squaring gives $a^{2^{\alpha-1}} = 1 + 2^{\alpha+1}t + 2^{2\alpha}t^2 \equiv 1 \pmod{2^{\alpha+1}}$. This implies that $a^{2^{\alpha-2}} \equiv 1 \pmod{2^\alpha}$ is always true for $\alpha \geq 3$, so, for each such α, the order of $a \pmod{2^\alpha}$ is at most $2^{\alpha-2} < \phi(2^\alpha)$.

(ii) $m = 2^{\alpha_0} p_1^{\alpha_1} \ldots p_r^{\alpha_r}$ ($\alpha_0 \geq 0$, $\alpha_i \geq 1$ for $1 \leq i \leq r$), where the p_i are odd primes with either $r \geq 2$ or $r = 1$ and $\alpha_0 \geq 2$. Here

$$\phi(m) = \phi(2^{\alpha_0}) \cdot \phi(p_1^{\alpha_1}) \ldots \phi(p_r^{\alpha_r}),$$

so
$$\tfrac{1}{2}\phi(m) = \phi(2^{\alpha_0})[\tfrac{1}{2}\phi(p_1^{\alpha_1}) \ldots \phi(p_r^{\alpha_r})]$$
$$= \phi(p_i^{\alpha_i})[\tfrac{1}{2}\phi(2^{\alpha_0}) \ldots \phi(p_{i-1}^{\alpha_{i-1}})\phi(p_{i+1}^{\alpha_{i+1}}) \ldots].$$

As

$$2 \mid \phi(p_j^{\alpha_j}) = p_j^{\alpha_j-1}(p_j - 1)$$

and, if $\alpha_0 \geq 2$, $2 \mid \phi(2^{\alpha_0}) = 2^{\alpha_0-1}$, we see that in each of the relevant cases $\tfrac{1}{2}\phi(m)$ is a common multiple of

$$\phi(2^{\alpha_0}), \ldots, \phi(p_r^{\alpha_r}).$$

If $\tfrac{1}{2}\phi(m) = K_0 \cdot \phi(2^{\alpha_0}) = K_i\phi(p_i^{\alpha_i})$ say, where the K's are integers, then for any a relatively prime to m,

$$a^{\frac{1}{2}\phi(m)} = (a^{\phi(2^{\alpha_0})})^{K_0} \equiv 1 \pmod{2^{\alpha_0}}$$

and for each i,

$$a^{\frac{1}{2}\phi(m)} = (a^{\phi(p_i^{\alpha_i})})^{K_i} \equiv 1 \pmod{p_i^{\alpha_i}}.$$

These congruences imply $a^{\frac{1}{2}\phi(m)} \equiv 1 \pmod m$ so that the order of a can be at most $\tfrac{1}{2}\phi(m) < \phi(m)$.

Answers to the Exercises

Chapter One

1. If $(m, n) = d$, then $d\left(\dfrac{m}{d} \cdot n\right) = mn$ where $\dfrac{m}{d} \cdot n = m \cdot \dfrac{n}{d}$ is a multiple of both m and n and is therefore a multiple of $[m, n]$. Thus $d \cdot [m, n]q = mn$ for some integer $q > 0$, so $dq \cdot \dfrac{[m, n]}{m} = n$ and $dq \dfrac{[m, n]}{n} = m$ implying that dq is a common divisor of m and n. Since d is the greatest common divisor, $q = 1$ and $d \cdot [m, n] = mn$.

2. (i) 1,   (ii) 3,   (iii) 1,   (iv) 7.

3.   (i) $1 = 14 \cdot 24 - 5 \cdot 67$,   (ii) $3 = 5 \cdot 15 - 72$,
     (iii) $1 = 26 \cdot 36 - 17 \cdot 55$,   (iv) $7 = 2 \cdot 126 - 7 \cdot 35$.

4. If $d = (a_1, \ldots, a_n)$ and $d_1 = (a_1, (a_2, \ldots, a_n))$, then $d_1 \mid a_1$ and $d_1 \mid (a_2, \ldots, a_n)$ so $d_1 \mid a_1$ and $d_1 \mid a_2, \ldots, d_1 \mid a_n$ whence $d_1 \leqslant d$. Suppose inductively that $(a_1, \ldots, a_k)$ is a linear combination of $a_1, \ldots, a_k$ for $k < n$ (true for $k = 2$). Then $d_1 = a_1 x_1 + y(a_2, \ldots, a_n) = a_1 x_1 + (a_2 x_2 + \ldots + a_n x_n)y$, is a linear combination of $a_1, \ldots, a_n$, whence $d \mid d_1$, so $d_1 \leqslant d$ gives $d = d_1$.

5. (i) 1;   (ii) 7;   (iii) 3;   (iv) 1.

6. If n is composite it has at least two prime divisors p_1, p_2 (perhaps equal). If $p_1 > \sqrt{n}$, $p_2 > \sqrt{n}$ then $p_1 p_2 > n$, a contradiction.

7. 2, 3, 5, 7, 11, 13, 17, 19, 23, 29, 31, 37, 41, 43, 47, 53, 59, 61, 67, 71, 73, 79, 83, 89, 97.

8. (i) $2 \cdot 5 \cdot 13$, (ii) $2 \cdot 3^2 \cdot 7 \cdot 11$, (iii) $13 \cdot 19$, (iv) prime, (v) $7 \cdot 11 \cdot 13$.

13. $\displaystyle\sum_{h \mid n} |\mu(h)|$ is a multiplicative function and $\displaystyle\sum_{h \mid p^r} |\mu(h)| = 1 + |\mu(p)| = 2$.

85

ANSWERS TO THE EXERCISES

Chapter Two

1. $N \equiv c_0 - c_1 + c_2 - \ldots + (-1)^n c_n \pmod{11}$.

2. Using $2^5 \equiv 2 \pmod{10}$ we have $2^{127} = 4 . (2^5)^{25} \equiv 4 . 2^{25} \equiv 4 . 2^5 \equiv 4 . 2 \equiv 8 \pmod{10}$, so $2^{127} - 1 \equiv 7 \pmod{10}$.

3. (i) congruence solution $x \equiv 1 \pmod 8$, equation solutions $x = -7, 17$;
   (ii) congruence has no solutions, so neither has equation;
   (iii) congruence $x \equiv 0, -3, 4 \pmod{13}$, equation $x = 13, 17$;
   (iv) congruence $x \equiv 2, 3 \pmod 5$, equation has no solutions.

4. (i) $x \equiv 5 \pmod 9$;   (ii) not soluble;   (iii) $x \equiv 5, 11, 17 \pmod{18}$;
   (iv) $x \equiv 4 \pmod 7$.

5. $\{1, 2, 3, 4\}$ and $\{1, -1\}$ are reduced sets of residues modulo 5 and 6 respectively. So a reduced set of residues modulo 30 consists of $5 . 1 + 6 . 1 = 11$, $5 . 1 + 6 . 2 = 17$, $5 . 1 + 6 . 3 = 23$, $5 . 1 + 6 . 4 = 29$, $5 . (-1) + 6 . 1 = 1$, $5 . (-1) + 6 . 2 = 7$, $5 . (-1) + 6 . 3 = 13$, $5 . (-1) + 6 . 4 = 19$.

6. If p is odd then $\phi(p) = p - 1 = \phi(2) . \phi(p) = \phi(2p)$; and if n is odd, then $\phi(2n) = \phi(2) . \phi(n) = \phi(n)$.

8. Both assertions are true if $\alpha = 3$. So, for $\alpha > 3$, assume inductively that 3 is of order $2^{\alpha-3} \pmod{2^{\alpha-1}}$ and that $3^{2^{\alpha-3}} = 1 + 2^{\alpha-1}t_{\alpha-1}$ where $2 \nmid t_{\alpha-1}$. Then $3^{2^{\alpha-2}} = (1 + 2^{\alpha-1}t_{\alpha-1})^2 = 1 + 2^{\alpha}t_{\alpha-1} + 2^{2\alpha-2}t_{\alpha-1}^2 = 1 + 2^{\alpha}(t_{\alpha-1} + 2^{\alpha-2}t_{\alpha-1}^2) = 1 + 2^{\alpha}t_{\alpha}$ where $2 \nmid t_{\alpha}$; and also the order d of 3 modulo 2^{α} must divide $2^{\alpha-2}$. But $3^{2^{\alpha-3}} = 1 + 2^{\alpha-1}t_{\alpha-1} \not\equiv 1 \pmod{2^{\alpha}}$ so that $d > 2^{\alpha-3}$. Hence $d = 2^{\alpha-2}$ as required.

9. (i) $x \equiv 23 \pmod{36}$;   (ii) $x \equiv -2 \pmod{105}$;
   (iii) $x \equiv 11, 31 \pmod{40}$, or equivalently $x \equiv 11 \pmod{20}$.

10. $x \equiv 2, 4 \pmod 5$.

11. $x \equiv 7, 43 \pmod{84}$.

Chapter Three

1. We know that $x^{p-1} - 1 \equiv 0 \pmod p$ has $p - 1$ solutions. Also $x^{p-1-d} + \ldots + 1 \equiv 0 \pmod p$ has at most $p - 1 - d$ solutions, so that $x^d - 1 \equiv 0$ must have at least d solutions (to bring the total number up to $p - 1$). But, being of degree d, $x^d - 1 \equiv 0$ cannot have more than d solutions, so it has exactly d.

2. If $aa' \equiv 1 \pmod p$ with $a = a'$ then $a^2 \equiv 1 \pmod p$ so that $a \equiv \pm 1 \pmod p$. Thus, apart from 1 and $p - 1$, the $p - 3$ numbers $2, \ldots, p - 2$ can be separated into $\frac{1}{2}(p - 3)$ pairs, $a_1, a_1'; a_2, a_2'; \ldots$ such that $a_i a_i' \equiv 1 \pmod p$ for each pair. Therefore

$$(p - 1)! = 1 . (p - 1) \prod_{i=1}^{\frac{1}{2}(p-3)} a_i a_i' \equiv 1 . (-1) . 1 \equiv -1 \pmod p.$$

3. n composite means $n = ab$ where $1 < a < n$, $1 < b < n$. If $a \neq b$, a and b both occur among $1, \ldots, n - 1$ so that $ab = n$ divides $(n - 1)!$ If $a = b$ and $n = a^2$ then, if $a > 2$, $a < 2a < n$ so that a and $2a$ both occur among $1, \ldots, n - 1$ whence $2a^2 = 2n$ divides $(n - 1)!$ Otherwise $n = 2^2$, which has already been mentioned.

5. (i) $x \equiv 8 \pmod{17}$; (ii) $x \equiv \pm 7 \pmod{17}$; (iii) $x \equiv 10 \pmod{17}$.

6. $p \equiv \pm 1 \pmod{12}$.

7. (i) 6 numbers; (ii) 22; (iii) 14.

8. (i) Yes, $x \equiv 12, 24 \pmod{41}$; (ii) No; (iii) Yes, $(399 = 3 . 7 . 19)$ and solutions are $x \equiv 28, 85, 161, 175, 218, 232, 308, 365 \pmod{399}$.

Chapter Four

1. (i) $6^2 + 31^2$, (ii) $3^2 + 9^2$, (iii) $10^2 + 11^2 = 5^2 + 14^2$.

2. $533 = (2 . 4 + 3 . 5)^2 + (2 . 5 - 3 . 4)^2 = 23^2 + 2^2 = (2 . 5 + 3 . 4)^2 + (2 . 4 - 3 . 5)^2 = 22^2 + 7^2$.

3. (i) $7^2 + 3^2 + 1^2 + 1^2 = 6^2 + 4^2 + 2^2 + 2^2 = 5^2 + 5^2 + 3^2 + 1^2$,
   (ii) $2^2 + 9^2 = 6^2 + 7^2$, (iii) $1^2 + 2^2 + 4^2$, (iv) 19^2,
   (v) $6^2 + 3^2 + 1^2 + 1^2 = 5^2 + 3^2 + 3^2 + 2^2$.

4. $2n + 1 = (n + 1)^2 - n^2$, $2n = (n + 1)^2 - n^2 - 1^2$. Any difference of two squares is $\equiv 0, \pm 1 \pmod 4$ so 2 requires three non-zero terms.

5. $8n + 3$ is certainly a sum of three squares and they must all be odd otherwise their sum would not be $\equiv 3 \pmod 8$.

6. $4(2k + 1) = 8k + 4 = 1 + (8k + 3) = 1^2 + (2n_1 + 1)^2 + (2n_2 + 1)^2 + (2n_3 + 1)^2$.

7. For any natural number a, $8a + 3 = (2n_1 + 1)^2 + (2n_2 + 1)^2 + (2n_3 + 1)^2$ which implies $8a = 4n_1^2 + 4n_1 + 4n_2^2 + 4n_2 + 4n_3^2 + 4n_3$, and so $a = \frac{1}{2}n_1(n_1 + 1) + \frac{1}{2}n_2(n_2 + 1) + \frac{1}{2}n_3(n_3 + 1)$.

8. $23 = 2 . 2^3 + 7 . 1^3, 239 = 5^3 + 3 . 3^3 + 4 . 2^3 + 1^3 = 2 . 4^3 + 4 . 3^3 + 3 . 1^3$.

9. When expressed as a sum of cubes, each of the infinitely many numbers $9m \pm 4$ requires at least four summands.

Index

39   1738   9